Thermodynamik

Energie • Umwelt • Technik

Band 20

λογος

Thermodynamik

Energie • Umwelt • Technik

Herausgegeben von Professor Dr.-Ing. Dieter Brüggemann
Ordinarius am Lehrstuhl für Technische Thermodynamik und
Transportprozesse (LTTT) der Universität Bayreuth

Combustion and In-Cylinder Soot Formation Characteristics of a Neat GTL-Fueled DI Diesel Engine

Von der Fakultät für Angewandte Naturwissenschaften

der Universität Bayreuth

zur Erlangung der Würde eines

Doktor-Ingenieurs (Dr.-Ing.)

genehmigte Dissertation

vorgelegt von

Salih Manasra, M.Sc.

aus

Haifa (Israel)

Erstgutachter: Prof. Dr.-Ing. D. Brüggemann

Zweitgutachter: Prof. em. Dr.-Ing. Dr.-Ing. E.h. mult. F. Mayinger

Tag der mündlichen Prüfung: 19. September 2011

Lehrstuhl für Technische Thermodynamik und Transportprozesse (LTTT)

Universität Bayreuth

2011

Vorwort des Herausgebers

Aus verschiedenen Gründen wird überlegt, ob man konventionellen Dieselkraftstoff nicht durch andere Kohlenwasserstoffe ersetzen könnte. Besonderes Interesse findet dabei das seit Jahrzehnten bekannte Fischer-Tropsch-Verfahren, durch das beispielsweise Erdgas in flüssige Kohlenwasserstoffe umgewandelt werden kann. So gewonnene sogenannte Gas-To-Liquid- oder GTL-Kraftstoffe weisen verbrennungstechnisch eine Reihe von Vorzügen auf. Hierzu zählen die hohe Cetanzahl, ein relativ niedriger Kohlenstoff- bzw. hoher Wasserstoffgehalt, das Fehlen von Schwefel und anderen unerwünschten Stoffen sowie die allgemeinen Vorteile von Kraftstoffen im flüssigen Aggregatzustand.

Es ist bekannt, dass GTL-Kraftstoffe in konventionellen Dieselmotoren eingesetzt werden können, ohne dass man deren Konstruktion nennenswert zu verändern braucht. Allerdings muss die Betriebsweise des Motors dem jeweiligen Kraftstoff angepasst werden. Dies gilt ganz besonders für die Einspritzung, die in heutigen modernen Dieselmotoren so ausgefeilt ist, dass sie eine nahezu vollständige Verbrennung mit sehr geringem Schadstoffausstoß und auch geringer Lärmentwicklung bewirkt. Zur Einspritzstrategie zählt nicht nur der Einspritzdruck und der Einspritzbeginn sondern der gesamte zeitliche Verlauf des Einspritzvorgangs.

Der Autor hat sich eingehend mit der Einspritzung und Verbrennung solcher Kraftstoffe beschäftigt und besonders die Rußbildung in GTL-betriebenen Dieselmotoren untersucht.

Bayreuth, im Oktober 2011 Professor Dr.-Ing. Dieter Brüggemann

PREFACE

This work was performed while serving as Research Associate at the Bayreuth Engine Research Center and the Institute for Thermodynamics and Transport Processes [Lehrstuhl für Thermodynamik und Transportprozesse] [LTTT] of the University of Bayreuth-Germany.

I am wholeheartedly thankful, therefore, for Prof. Dr.-Ing Dieter Brueggemann for supervising this work and all the support he has kindly offered.

I am also wholeheartedly grateful for Prof.em. Dr.-Ing Dr.-Ing E.h. Franz Mayinger for his insightful advices and sincere support throughout the whole period of my engagement as Research Associate at both, the Institute for Thermodynamics and Transport Processes of the University of Bayreuth and the Institute of Thermodynamics of the Technical University of Munich-Germany.

My special appreciation I owe to Prof. Dr.-Ing Frank Rieg for presiding the Doctoral Committee and overtaking all due duties.

It is needless to assert my great gratitude to all the colleagues, technicians, secretaries and staff of both the Bayreuth Engine Research Center and the Institute for Thermodynamics and Transport Processes of the University of Bayreuth for all the support and friendly engagement they have kindly offered.

I also wholeheartedly thank, Mr. Azman Abd-Rashid of SHELL London and Mr. Rcihard Clark, Mr. Felix Balthasar and Dr. Dorothea Liebig of SHELL Global Solutions [SGS] for providing this work with SHELL's GTL-Fuel and for all the related technical support they have kindly offered.

Last but not least, I, with great gratitude, dedicate this work with all pertaining rights to my dear and beloved mother for all the support and motivation she has given me.

(“*Allah (god) will exalt those who believe among you,*
and those who have knowledge, to high ranks.
Allah is Informed of what ye do.”)
The Holy Qur'an

ABSTRACT

In the light of the increasing stringent pollutant emissions regulations, alternative fuels are emerging as more promising solutions for meeting future emission standards. Natural gas although superior in terms of fuel properties seems to have limited use in transport vehicles due to storage and infrastructure-related issues as well as engine modification and adaptation. Gas-To-Liquid [GTL] fuel which is basically the end product of the Fischer-Tropsch process in which synthesis gas derived from natural gas is converted into liquid hydrocarbon, seems, as a result to become the optimal alternative fuel of choice. It bears fuel properties quite close to natural gas namely high cetane number, low Carbon-to-Hydrogen ratio, zero sulfur and near-zero aromatic content while on the other hand maintaining all characteristics of liquid fuels. Thus, GTL fuel can be used in existing diesel engines with, uniquely, minimal engine modification requirements.

This work capitalizes on the utilization of GTL fuel in diesel combustion system and unlike the vast majority of investigations conducted up to now, focuses rather on fundamental studies of the in-cylinder processes. It deploys, therefore, an optically-accessed Rapid Compression Machine allowing the application of several optical diagnostics, namely, shadowgraph imaging for the visualization of in-cylinder spray development, imaging and non-imaging cycle-resolved Soot Incandescence as well as OH-Chemiluminescence imaging. It seeks to investigate the effects of physical parameters, namely compression temperature and pressure, engine parameters, namely injection pressure and timing as well as the effect of injection strategy and rate shaping, namely single injection versus split and multiple injection on the in-cylinder combustion and soot formation characteristics. Emphasis is placed, however, on the type of combustion modes which can be realized through the utilization of GTL fuel.

It has been shown that lowering compression temperature by decreasing Compression Ratio [CR] to 14 prolonged the ignition delay of GTL fuel and permitted the realization of Low CR premixed combustion. Increasing compression pressure, however, yielded higher soot formation rate while increasing injection pressure brought about reduction in both soot formation rate and peak. The effect of both compression and injection pressures on soot formation seem to be related to their effect on Lift-Off-Length [LOL]. Retarding injection timing allowed an effective utilization of cooling effect improving mixing and yielding lowest soot level. The combination of late injection and low CR allowed the realization of HPLI combustion mode. Split injection on the other hand had altered combustion mode from fully pre-mixed to a combined pre and non-premixed combustion mode. The time between the two injection events had strong influence on combustion and soot formation. Surprisingly, in the case of GTL, this time doesn't affect ignition delay of the main injection and shortest time yielded lowest soot level and highest heat release peak. Here as well, the effect of this time on soot formation tends to be related to LOL. Post injection, regardless of timing, seems to produce additional amount of soot but early post injection yielded highest soot oxidation rate because of higher OH concentration and higher mixing energy.

Key words: GTL combustion, Low CR Premixed Diesel Combustion, HPLI Combustion, Lift-Off-Length, Split Injection; Multiple Injection, Post Injection, Shadowgraph, Soot Incandescence, OH-Chemiluminescence.

KURZFASSUNG

Angesichts der steigenden strengen Abgasbestimmungen tauchen alternative Brennstoffe als hoffnungsvolle Lösungen auf, um die zukünftigen Abgasnormen erfüllen zu können. Trotz der überragenden Brennstoffeigenschaften von Erdgas scheint dieser Brennstoff nur einen begrenzten Einsatz bei Transportfahrzeugen zu finden, was sich in der Lagerung und infrastrukturellen Problemen sowie der Motormodifikation und Anpassung begründet. Gas-To-Liquid [GTL] Brennstoff, generelles Endprodukt des Fischer-Tropsch Verfahrens, bei welchem Synthesegas, das aus Erdgas gewonnen wird in flüssigen Kohlenwasserstoff umgewandelt wird, scheint hier der optimale alternative Auswahlbrennstoff zu sein. GTL hat Brennstoffeigenschaften, die nahe an denen des Erdgases liegen, in diesem Falle eine hohe Cetanzahl, ein niedriges Kohlenstoff zu Wasserstoff Verhältnis, Null Schwefel und einen Aromatengehalt von nahezu Null und es behält dennoch alle Eigenschaften von flüssigen Brennstoffen. Daher kann der GTL Brennstoff in bereits existierenden Dieselmotoren eingesetzt werden, wobei hier nur ein geringfügiger Bedarf für eine Modifikation am Motor besteht.

Diese Ausarbeitung schlägt ihr Kapital aus dem Einsatz von GTL Brennstoff in Dieselmotorsystemen und, anders als beim größten Teil der bisher durchgeführten Untersuchungen, konzentriert sich diese Ausarbeitung mehr auf die Grundlagenstudien über die innermotorischen Prozesse. Es wird deshalb eine computergesteuerte schnelle Kompressionsmaschine eingesetzt, die den Einsatz von mehreren optischen Diagnosen, nämlich das Schattenriss Verfahren zur Darstellung des innermotorischen Sprayverlaufs, abbildende und nichtabbildende zeitaufgelöstes Ruß-Eigenleuchten sowie die OH-Chemilumineszenz Darstellung erlaubt. Es werden die Effekte der physikalischen Parameter wie beispielsweise Kompressionstemperatur und Kompressionsdruck, die Motorparameter wie beispielsweise der Einspritzdruck und Timing sowie der Effekt der Einspritzstrategie und Einspritzverlauf, hier die Einzeleinspritzung gegenüber der Split-Einspritzung und der Mehrfacheinspritzung bei der innermotorischen Verbrennung und die Rußbildungseigenschaften untersucht. Der Schwerpunkt ist jedoch auf die Verbrennungsverfahren ausgerichtet, die durch den Einsatz von GTL Brennstoff realisiert werden können.

Es hat sich gezeigt, das eine Absenkung der Kompressionstemperatur durch Minderung des Verdichtungsverhältnisses [VV] auf 14, den Zündverzug des GTL Brennstoffes verlängert und so die Realisierung einer vorgemischten Verbrennung mit einem Niedrig-Verdichtungsverhältnis erlaubt. Die Erhöhung des Kompressionsdruckes ergab jedoch eine höhere Rußbildungsrate aber die Erhöhung des Einspritzdruckes ergab eine Reduktion der Rußbildungsrate und -spitze. Der Einfluss von Kompressions und Einspritzdruck auf die Rußbildung scheint eng verbunden zu sein mit ihrem Einfluss auf die Lift-Off-Length [LOL]. Eine Verzögerung der Einspritz-Timing erlaubt den effektiven Einsatz eines kühlungsförderlichen Gemisches mit gleichzeitig niedrigstem Ruß Grad. Die Kombination von verspäteter Einspritzung und niedrigem Verdichtungsverhältnis erlaubt die Realisation der HPLI Verbrennung. Eine Split-Einspritzung auf der anderen Seite veränderte die Verbrennung von komplett vorgemischter Verbrennung zu einer kombinierten vorgemischten und nicht vorgemischten Verbrennung. Die Zeit zwischen zwei Einspritzungen hatte einen starken Einfluss auf die Verbrennung und die Rußbildung. Überraschenderweise, im Fall von GTL Brennstoff, hat diese Zeit keinen Einfluss auf den Zündverzug der Haupteinspritzung und die kürzeste Zeit erzielte den geringsten Ruß Grad und die höchste Verbrennungsverlaufspitze. Auch hier scheint der Einfluss dieser Zeit auf die Rußbildung eng mit der der Lift-Off-Length [LOL] verbunden zu sein. Ungeachtet von der Timing, scheint eine Nacheinspritzung eine zusätzliche Rußbildung zu produzieren, eine frühe Nacheinspritzung ergab jedoch die höchste Rußoxidationsrate wegen der höheren OH Konzentration und einer höheren Mischenergie.

Schlagwörter: GTL Verbrennung, Vorgemischte Dieselverbrennung mit niedrigem Verdichtungsverhältnis, HPLI Verbrennung, Lift-Off-Length, Split Einspritzung; Mehrfacheinspritzung, Nacheinspritzung, Schattenriss, Ruß-Eigenleuchten, OH-Chemiluminesz

Contents

NOMENCLATURE

γ	Ratio of specific heats	[-]
ε_x	Displacement in x-direction	[m]
ε_y	Displacement in y-direction	[m]
θ	Mirror's Tilting angle	[°]
ρ	Density	$[Kg/m^3]$
τ_{id}	Ignition delay	[s]
τ_{flow}	Characteristic flow time	[s]
τ_{chem}	Characteristic chemical time	[s]
ϕ	Equivalence ratio	[-]
$\boldsymbol{A}$	Constant	
$\boldsymbol{a}$	Cut-off ratio	[%]
A_n	Total orifice area	[m^2]
AR	Area reduction of a nozzle	[-]
$\boldsymbol{C}$	Contrast threshold	
C_D	Orifice flow rate (discharge) coefficient	
C_p	Specific heat at constant presure	[J/Kg·K]
C_V	Specific heat at constant volume	[J/Kg·K]
$\boldsymbol{Da}$	Damkoehler number	[-]
D_D	Droplet diameter	[m]
D_{inlet}	Nozzle orifice inlet diameter	[m]
D_{outlet}	Nozzle orifice outlet diameter	[m]
$\boldsymbol{E_A}$	Apparent activation energy	[J]
$\boldsymbol{f}$	Mirror focal point	
K	Conicity factor	[m]
L	Schlieren extent	[m]
$\boldsymbol{M}$	Molecular mass	[Kg]
$\boldsymbol{n}$	Exponential constant	
n_0	Refraction index	
$\boldsymbol{P}$	Pressure	[Pa]
$\mathbf{P_{CH}}$	Charge pressure (Boost pressure)	[Pa]/[bar]
$\mathbf{P_{INJ}}$	Injection pressure	[Pa]/[bar]
$\mathbf{P_{SOI}}$	Pressure at Start Of Injection	[Pa]/[bar]
Q_{ch}	Gross heat release	[J]
Q_{ht}	Heat transferred to the walls	[J]
Q_n	Net heat release	[J]
$\tilde{R}$	Universal (ideal=gas constant)	[J/mol·K]
Ra	Inlet rounding radius	[m]
Re	Reynolds number	

Re_l	Liquid Reynolds Number	[-]
S	Jet penetration length	[m]
T	Temperature	[K]
t_{br}	Time for jet break-up	[s]
T_{SOI}	Temperature at Start Of Injection	[K]
V	Volume	[m^3]
V_{rel}	Relative velocity	[m/s]
We	Weber Nuber	[-]

Abbreviations

ATDC	After Top Dead Center	
BDC	Bottom Dead Center	
BSFC	Brake-Specific Fuel Consumption	
BTDC	Before Top Dead Center	
BTE	Brake Thermal Efficiency	
CA	Crank Angle	
CCD	Charge-Coupled Device	
CN	Cetane Number	
CR	Compression Ratio	
DAS	Data Acquisition System	
DI	Direct Injection	
EGR	Exhaust Gas Circulation	
FT	Fischer-Tropsch	
GTL	Gas To Liquid	
HC	HydroCarbons	
HCCI	Homogenous Charge Compression Ignition	
HPLI	Highly Premixed Late Injection	
HPP	High Pressure Pump	
HRR	Heat Release Rate	
HS	High-Speed	
ICCD	Intensified Charge-Coupled Device	
KN	Cavitation Number	
LCRP	Low CR Premixed	
LHV	Lower Heating Value	[J/Kg]
LIF	Laser-Induced Fluorescence	
LII	Laser-Induced Incandescence	
LOL	Lift-Off Length	[m]
LTC	Low Temperature Combustion	
MI	Multiple Injection	
MK	Modulated Kinetics	
NOx	Nitrogen oxides	
PAH	Polycyclic Aromatic Hydrocarbons	
PLIF	Planer Laser-Induced Fluorescence	
PM	Particulate Matter	
RCM	Rapid Compression Machine	
ROHR	Rate Of Heat Release	

SI	Split Injection
SINL	Spatially Integrated Natural Luminosity
SISI	Spatially Integrated Soot Incandescence
SMI	Single Main Injection
SOI	Start of Injection
Syngas	Synthesis gas
TDC	Top Dead Center
UHC	Unburned HydroCarbons
VCO	Valve Covered Orifice

1 Introduction

Diesel engines would continue to serve as power source in both light and heavy duty vehicles for the days to come. It is superior over other engine types in many aspects. The growing demands for cuts in fuel consumption and exhaust emissions necessitate, however, extraordinary measures and alternative powering methodologies. New and innovative developments in diesel engine technologies have emerged as a result ranging from advance and alternative combustion concepts through the utilization of advance fuel injection strategies and systems to advance exhaust after treatment systems. These advancements alone don't bear, however, the capability to meet the more stringent future emission regulations and great efforts are being invested to come up with practical, yet cost-effective solutions. The utilization of alternative fuel is emerging on one hand as a non avoidable measure while new powertrain concepts seem to be inevitable. Interestingly, however, is that promising new powertrain concepts, namely parallel or series hybrid-electric, still use diesel engine in one form or another, as prime mover as well as electric generator known otherwise as range extender. It is sought, therefore, to have diesel engine, whether in conventional or advance powertrain systems operated on alternative fuels so to further improve emissions and fuel consumption performances. Biodiesel has been considered as good alternative diesel fuel and federal regulations have already set policies to have diesel engines operated on blends of biodiesel. The additional costs and necessary engine modification work associated with the operation on biodiesel-based fuels seem not to be very motivating. Other fuels, namely Gas-To-Liquid and Biomass-To-Liquid fuels, are emerging as more attractive solutions. They can be used either as blends or simply neat fuels and require almost no modification in engines and in supplying infrastructure either. Their performance is superior to biodiesel in many aspects. Lowe fuel consumption, lower exhaust emissions, lower maintenance and retail cost as well as lower engine noise. Despite the relatively high production cost of GTL and BTL many oil companies are not hesitating in investing in GTL production projects given their promising performance and few of them have reached large-scale commercial production plants, the largest of which is the SHELL GTL PEARL production plant in Qatar.
Large volume of research work has been invested to establish an evaluation of the performances of GTL-fueled diesel engines. They range from laboratory testing to field experimentation. The most recent testing by TNO of the Netherlands of urban buses fueled with neat GTL, neat biodiesel and CNG showed that GTL brings about the best reduction rates in both PM and NOx emission as well as operation costs for all vehicle classes, Euro 3, Euro 4 and Euro 5. Only CNG could yield lesser NOx emissions rates but from the cost-effectiveness point of view, GTL is more superior.

The superior performance of GTL fuel continues to attract more research work. The vast majority of the work is, however, practical and application-oriented. They are often comparative in nature, comparing the performance of several alternative fuels with GTL for a given engine or vehicle type.

This work is, however, fundamental-oriented and focuses on the investigation of the effects of physical, engine and injection parameters on the combustion and soot formation characteristics of direct injection diesel engine fueled with neat GTL fuel.

It focuses on the in-cylinder process to establish an understanding of the physical aspects of GTL fuel combustion. It deploys, therefore, an optically-accessed rapid compression machine and applies several optical diagnostics. It seeks to find out how key physical parameters such as cylinder (compression) temperature and pressure, key engine parameters namely injection pressure and timing, and injection rate shaping affect the in-cylinder soot formation and combustion processes of neat GTL-fueled diesel engine. It aims also at evaluating how the already established knowledge for conventional diesel fuel combustion in the context of the investigated parameters are preserved or differ when using GTL fuel.

Furthermore, given that emphasis is placed on advance or alternative combustion concepts which are realized through the combination of non-conventional injection and combustion methodologies with alternative fuel, this work aims, also, at providing initial assessment of the potential of the utilization of GTL fuel in the realization of non-conventional combustion modes and methods such as those of premixed-oriented diesel combustion.

2 Theoretical Background and Literature Assessment

2.1 Combustion in Diesel Engines

Diesel engines have been widely used for more than a century since Rudolf Diesel's invention of the compression ignition engine in 1898. Its numerous advantages, mainly higher efficiency and lower fuel consumption, make it a favorable engine for many of many applications. Despite the tremendous knowledge acquired in understanding diesel combustion engine throughout the past century, combustion in diesel engines is still a complex and not fully understood phenomenon. This can be attributed to the many various mechanical, physical and chemical processes it undergoes while each and every one of it is by itself a complex phenomenon.

To better understand combustion in diesel engines one needs to consider the various processes associated with it and analyze the role of each in the overall combustion process. First, fuel is pressurized within the injection system and is introduced into the cylinder chamber of a diesel engine through a nozzle with injection pressure up to 2000 bars [1], [2] (modern diesel engines). The injected fuel jet enters the chamber at sufficiently high velocities to atomize into small-sized droplets to enable rapid evaporation and traverses the combustion chamber in a cone-shaped spray within the time available [3]. During this time, the droplets formed evaporate, mix with the surrounding air charge and burn upon reaching the flammability limit. The time between start of injection until burning begins is called ignition delay and consists of two main elements: 1) Physical delay which corresponds to the time through which the fuel jet atomizes and breaks up into droplets allowing fuel evaporation and later mixing with air. 2) Chemical delay which corresponds to the time delay until the pre-combustion chemical reactions take place leading to auto ignition. This brief description clearly exposes the complexity of combustion in diesel engine and elaboration of each process of the above will now be demonstrated for further clarification.

Fuel Injection and Spray Overall Structure

Following fuel pressurization within the injection system fuel, upon injector's needle lift, flows through the injector nozzle to enter the combustion chamber. The fuel internal flow within the nozzle undergoes depending on nozzle geometry and other parameters that would be discussed in more details in the subsequent sections cavitating process mainly because of fuel throttling and flow turning while passing through the nozzle hole. This leads to flow acceleration along the nozzle turning edge which in return drops the local pressure to pressures below vapor pressure leading to the emergence of cavitation bubbles [4] and flow separation along the nozzle hole's walls [5]. The occurrence of cavitation can lead first to the reduction of the effective cross sectional area of the fluid and to turbulence intensification within the flow to a degree that can cause an early spray break-up [6]. Figure 2-1 demonstrates this process qualitatively. As the liquid jet leaves the nozzle it becomes turbulent and penetrates the combustion chamber as it entrains and mixes with the surrounding air.

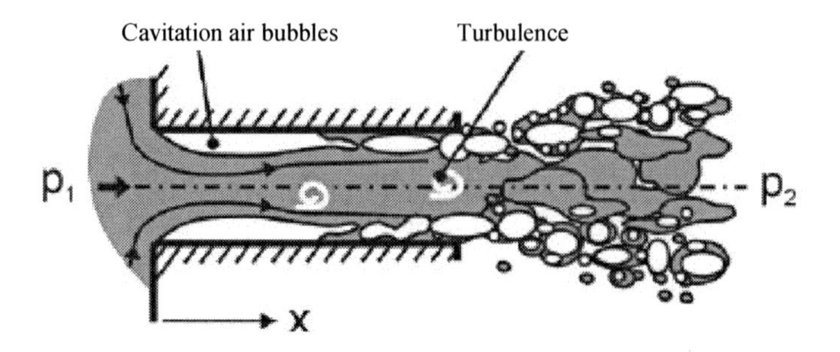

Figure 2-1: Flow structure through a nozzle hole upon emergence of cavitation. From Baumgarten [6]

The outer surface of the jet breaks up into drops of order of 10 µm diameter close to nozzle exit [3]. This liquid column leaving the nozzle disintegrates over the finite length called the *breakup length.* Each liquid fuel jet atomizes into drops and ligaments. While moving away from the nozzle, the mass of air within the spray increases, the spray diverges, its width increases and the velocity decreases. The highest velocities are on the jet axis while the equivalence ratio (fuel-to-air ratio) is highest on the centerline decreasing to zero (unmixed air) at the spray boundary. Figure 2-2 schematically shows the equivalence ratio (Φ) distribution within the fuel jet.

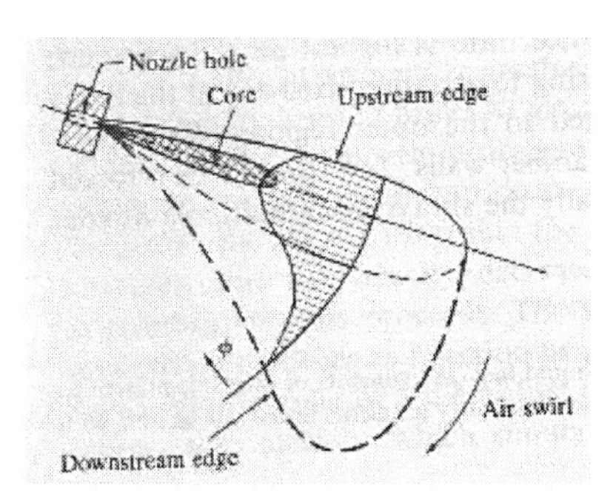

Figure 2-2: Schematic of fuel spray injected radially outward from the chamber axis into swirling air flow. Shape of equivalence ratio (ϕ) distribution within jet is indicated. From Heywood [3]

The droplets on the outer edge of the spray evaporate first, creating a fuel vapor-air mixture sheath around the liquid-containing core while the tip of the spray penetrates further into the combustion chamber as injection proceeds, but at a decreasing rate.
This is mainly because drops at the tip of the spray meet the highest aerodynamic resistance and slow down, but the spray continues to penetrate the air charge because when droplets retard at the tip they are continuously replaced by new higher-momentum later-injected drops [7]. Once the sprays have penetrated to the outer region of the combustion chamber they interact with the chamber wall.

The spray is then forced to flow tangentially along the wall (eventually the sprays from multi-hole nozzles interact with each other).
This spray-wall interaction could follow wall impingement, reflection and flow turning improve as well as deteriorate the quality of mixture formation depending on its influence on spray atomization and fuel spread out. Figure 2-3 demonstrates schematically the overall spray structure and break-up while figures 2-4 and 2-5 show spray images taken in this work using the Schlieren technique.

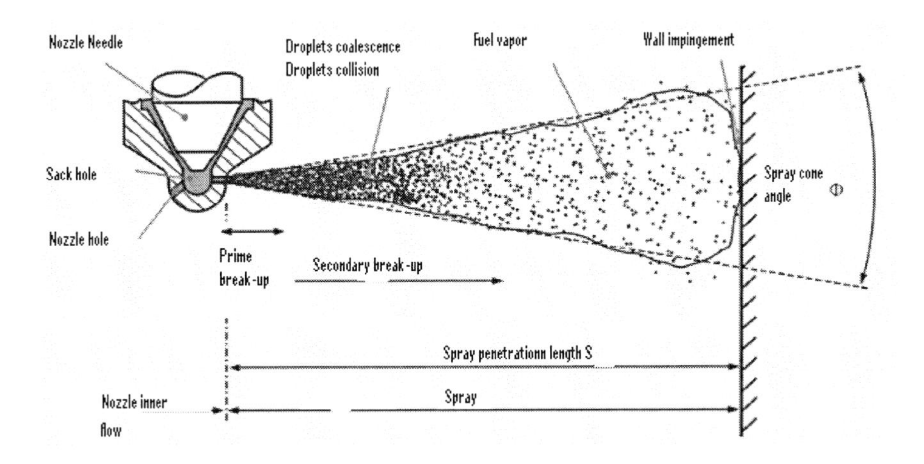

Figure 2-3: Schematic presentation of diesel spray structure. From Baumgarten [6]

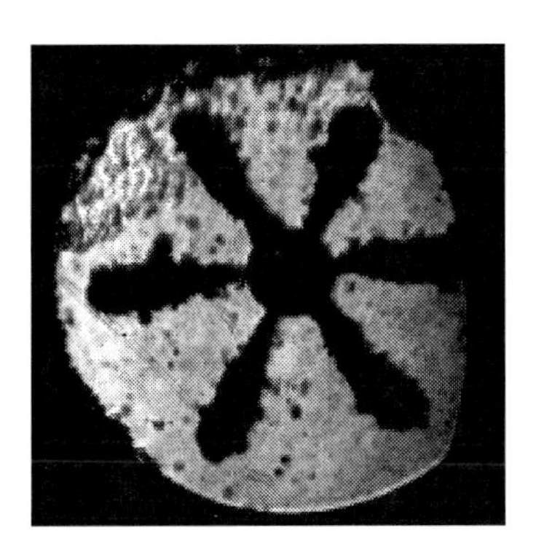

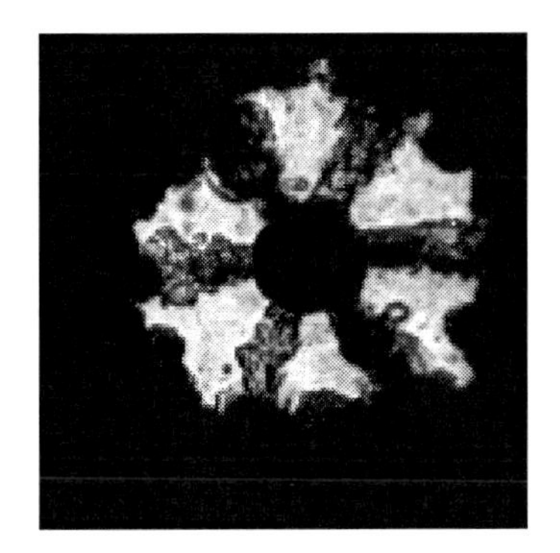

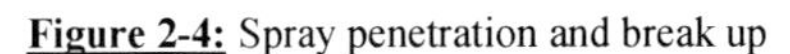

Figure 2-4: Spray penetration and break up

Figure 2-5: Fuel evaporation within liquid jet.

Three typical characteristics are used to describe spray development and fuel-air mixing quality. These are 1) spray penetration length, 2) spray cone angle and 3) droplet size. The inter-relation between these characteristics and spray development shall be demonstrated in the subsequent sub-sections, but first light will be shed on additional important processes.

Spray Break-up and Atomization

As depicted in figure 2-3, under typical diesel engine injection conditions, the fuel jet undergoes a two-stage atomization following entrance into the combustion chamber. In the area within the vicinity of the nozzle, break-up is due to the unstable growth of surface waves caused by surface tension and results in drops larger than the jet diameter. This is typically referred to as the *prime break-up* or the *atomization break-up regime*. Cavitating and turbulent flow right at the nozzle exit are responsible together with density and velocity differences between the liquid jet and surrounding air for the occurrence of such instabilities leading to first break-up. As injection continues the jet velocity increases and forces due to the relative motion of jet and surrounding air augment the surface tension force leading to a drop sizes of the order of the jet diameter. Further increase in jet velocity increases moment exchange between the liquid and the surrounding air resulting in second break-up in the *atomization regime* typically referred to as *secondary break up*. Depending on the surrounding air temperature, heat is transferred to the smaller drops initiated within the secondary break-up and evaporate.
Aerodynamic interaction at the liquid gas interface seems to be one major component of the atomization mechanism in the secondary break-up regime. This is usually described by the *Weber-Number* which expresses the dimensionless relation of the aerodynamic and surface tension forces. Or in other words it can be thought of as a measure of the relative importance of fluid's inertia and its surface tension. Weber-number is given as:

$$\mathrm{We_g} = \frac{\rho_g \cdot D_D \cdot \mathrm{v}_{\mathrm{rel}}^2}{\sigma} \qquad (2\text{-}1)$$

Where: D_D is the drop diameter and σ is the surface tension

In this connection Weber-number equal or greater than 1 determines droplets break-up and its capability to entrain the surrounding air [8]. A criterion for the onset of jet atomization at the nozzle exit plane was also developed. This is expressed in terms of Weber and Reynolds numbers as follows:

$$\left(\frac{\rho_l}{\rho_g}\right)\left(\frac{\mathrm{Re}_l}{We_l}\right)^2 > 1 \qquad (2\text{-}2)$$

Spray Evaporation

Another important factor which together with jet break-up and atomization make up the physical delay component of the ignition delay is spray or droplet evaporation as already mentioned. This is a very important yet complex process that influences not only ignition delay but also pollutant formation. Droplet evaporation is a critical process given the fact that air-fuel mixing can only take place while both components are in the gaseous phase. This is the last stage preceding air-fuel mixing and upon reaching flammability limit allowing ignition to take place.

Temperature of the surrounding air and the diffusion of the fuel along the droplet outer surface play a remarkable role in the evaporation process. Momentum relation between fluid and surrounding air is, however, a decisive factor for the interaction between air and fuel jet [5]. The process of droplet evaporation under normal diesel engine operating conditions seems to be rapid relative to the total combustion period [9] and undergoes under these conditions the following phenomena:

1. Deceleration of the drop due to aerodynamic drag
2. Heat transfer to the drop from the surrounding air
3. Mass transfer of vaporized fuel away from the drop to the surrounding air

As the droplet temperature due to heat transfer increases, the fuel vapor pressure increases and the evaporation rate increases while as the drop velocity decreases, the convective heat-transfer coefficient between the air and drop decreases [3]. Injection pressure, therefore, has a substantial effect on droplet evaporation and fuel-air mixing.

In the spray core, where drop number densities are high, the evaporation process has a significant effect on the temperature and fuel-vapor concentration in the air within the spray. As fuel vaporizes, the local air temperature decreases and the local fuel vapor pressure increases. This fact that liquid fuel vaporization causes significant reduction in gas temperature emphasizes the effect of chamber air temperature on the fuel evaporation rate and subsequently on the ignition delay, both physical and chemical delays.

There are enormous studies on droplet evaporation intending to establish models simulating it and the key conclusions from such studies are essentially that under normal diesel engine operating conditions, 70 to 90 percent of the injected fuel is in the vapor phase at the start of combustion. However, only 10 to 35 percent of the vaporized fuel has mixed to within the flammability limits in a typical medium-speed DI diesel engine [3]. Combustion is, therefore, largely mixing-limited rather than evaporation-limited.

Ignition Delay

Ignition in diesel engines marks the onset of combustion. Unlike in gasoline engines ignition is not directly controlled. It's characterized by *ignition delay* which as already mentioned is the time between the start of injection and the onset of mixture burning. The later is typically recognized by sudden temperature or pressure rise as well as by the increase in the concentration of free radicals or by the appearance of flame [5].
Actually experience has shown that under normal conditions, the point of appearance of flame is later than the point of pressure rise [3]. This ignition delay corresponds to the physical ignition delay which encompasses the various processes mentioned above from fuel atomization, evaporation and mixing with air leading to an ignitable state and to the chemical ignition delay through which series of chemical reactions of fuel oxidation and the build-up of free radicals take place allowing the initiation of the ignition chemical reaction. It is clear then that ignition delay depends on many factors, those which have direct effect on both the physical and chemical delays. A large number of studies have been conducted in attempt to come up with mathematical correlations for describing or quantifying ignition delay.

These turned to be of the form of empirical equations deduced from experimental data. Several forms have been proposed but all revolve around a common form:

$$\tau_{id} = Ap^{-n} \exp\left(\frac{E_A}{\tilde{R}T}\right) \tag{2-3}$$

Where τ_{id} is the ignition delay, E_A is an apparent activation energy for the fuel auto-ignition process, and A and n are constants dependent on fuel (and, to some extent, the injection and air-flow characteristics). From the above it's obvious that while the influence of the physical process are somewhat inherited in the empirical constants temperature and pressure effects dominate ignition delay. This can be attributed to the fact that ignition in diesel combustion is matter of fact a compression ignition hence a thermal-chemical ignition.
Practically several expressions have been proposed for ignition delay but the most common one is that of Watson [10] which is of the form:

$$\tau_{id}(ms) = 3.5(p/bar)^{-1.022} \exp\left(\frac{2100\,K}{T}\right) \tag{2-4}$$

Assanis et al. [11] have extended this expression to also incorporate the effect of mixture formation depicted by the equivalence ratio. Thus equation 2-4 is modified to be of the form:

$$\tau_{id}(ms) = 2.4\phi^{-02}(p/bar)^{-1.02} \exp\left(\frac{2100\,\mathrm{K}}{T}\right) \tag{2-5}$$

Although the influence of fuel type is somewhat inherited in the apparent activation energy some researchers have extended the expression for ignition delay to also include key fuel property such Cetane Number [CN]. These would be further elaborated in a subsequent section.

<u>Combustion Development and Heat Release</u>

It is obvious from the already mentioned processes that diesel combustion undergoes many processes and burning takes place in different modes. Combustion is better understood by looking at the heat release rate relative to the fuel injection rate. This describes the rate at which burning takes place and its correlation to pressure and temperature increase within the cylinder which can be used to interpret the development of the various physical and chemical processes.
A typical heat release rate shape is shown in figure 2-6 in which the following combustion stages are depicted:

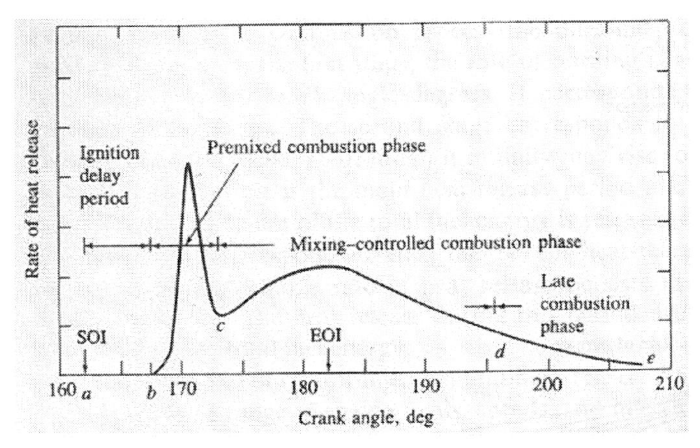

Figure 2-6: Typical DI engine heat-release-rate diagram identifying different diesel combustion phases. From Heywood [3]

Ignition delay (ab). This is the time between start of injection and the onset of combustion already explained in the preceding subsection. It's recognized by the initial positive rise of the heat release slope, hence initial positive gradient or optically by chemiluminescence emission. Studies by Dec and Espey [13] and Dec et al [14] showed that chemiluminescence arises mainly from formaldehyde (Ch_2O) and CH emission with no OH emission which is typical emissions for cool chemistry reactions of "cool flame".

Premixed or rapid Combustion Phase (bc). For combustion to take place mixture should have arrived the so called "flammability limit". In this stage the injected fuel that has been already evaporated and mixed with air to within the flammability limit burns rapidly in the form of premixed, relatively homogeneous combustion characterized by a high sudden pressure increase and thus steep heat release rate. This often leads to the so called "diesel nock" or "hard combustion" or "combustion noise". This is mainly because combustion in this regime occurs in the form of thermal explosion as compression brings the already mixed fuel with air to ignition. The fuel mass that burns in this combustion phase is linearly related to the product of engine speed and ignition delay time [12].
Higher compression temperatures, therefore, decreases the fuel mass burned in a premixed mode given its direct effect on shortening ignition delay. Flame luminosity in this combustion mode is relatively too low given its lower temperature and burning takes place in the outer envelop of the fuel spray mainly in the leading portion of the jet which has already entertained in air.

Mixing-Controlled Combustion (Diffusion Combustion) (cd). This is the second phase of combustion often related to as "main combustion" which starts once pressure rise in the premixed combustion has reached its peak and ends once the cylinder temperature has reached its maximum temperature. As soon as the premixed fuel-air mixture has been consumed in the premixed combustion, combustion becomes controlled by the rate at which fuel is made ignitable. Hence by the rate at which fuel atomizes, evaporates and mixes with the available air to become combustible.

Following the onset of combustion in the premixed phase the later injected fuel breaks up in smaller droplets and the heat transferred from the surrounding flame accelerates droplet evaporation but by that time a thin diffusion flame front has already been formed at the spray envelop to which reacting fuel droplet is transferred by the further fuel injection from one side and surrounding air from the other side forming a turbulent diffusion flame type burning. In this regime oxidation of partially burnt hydrocarbons, particles and carbon monoxides into water and CO_2 takes place. Given that mixing is the slowest process, combustion is indeed mixing-controlled. A dimensionless parameter called *Damkoehler number* is recalled to demonstrate the relative weight of mixing parameters which is flow-related parameters to the chemical parameters. It is expressed in the form of

$$Da = \frac{\text{characteristic flow time}}{\text{characteristic chemical time}} = \frac{\tau_{flow}}{\tau_{chem}} \tag{2-6}$$

Plee and Ahmad [15] calculated Damkoehler number for various diesel flames and showed that the time required to mix fuel and air is 500-2300 times longer than the time required to burn the resulting fuel-air mixture.
The vast majority of the injected fuel, usually more than 75 percent [3], burns in this combustion phase as a result the biggest portion of the heat release takes place in this phase during which yielding the largest quantities of combustion and pollutant products. Mixing-controlled combustion phase is indeed the decisive phase of the entire diesel combustion and determines diesel engine performance.

Late combustion phase (de). Heat release continues at a lower rate well into the expansion stroke. This third phase takes place from the point at which the maximum temperature has reached until the end of combustion. A small fraction of unburned fuel as well as fraction of the fuel energy that is still present in the form of soot alongside of fuel-rich combustion products can still burn releasing further heat. However due to pressure and temperature decrease as a result of piston motion backward towards bottom dead center chemical kinetics becomes really slow compared to mixing. Combustion in this phase is unlike the second phase reaction kinetics-controlled combustion in which depending on temperature, availability of oxygen and mixing, large portion of the already formed soot burns out through the oxidation reaction involving attack of OH radicals.
A trade-off between cylinder pressure, oxygen and radical concentration determines the final levels of pollutants' formation. Given the dependency of soot oxidation and NOx formation on temperature the development of combustion processes should aim at a combustion mode in which the evolution of temperature together with other influential measures are controlled in a such that either soot formed in the main combustion is too low or soot oxidation in the late combustion is too high or even both.

2.2 Soot Formation in Diesel Engines

Soot emission has always been the focal point of pollutant emission since it compromises the largest portion of particulate matter, in fact, over 50% of the emitted particulate matter is soot [16], and its control often conflicts with other major toxic pollutants such as NOx. Soot formation in diesel engines actually continues to interest researchers firstly due to the multi-stage it undergoes and secondly, its complexity. Despite major progression and significant knowledge acquired, soot is yet to be fully understood. The harsh environment in diesel engines and its multi-regime combustion only add order of magnitude to the complexity of the phenomenon.

Basically, soot is formed in fuel-rich zones and at relatively high temperatures such as temperatures above 1400 K [17] or above 1500 K [1]. It is formed, therefore, in both the fuel-rich premixed combustion as well as in the non-premixed diffusion combustion. Following its inception, soot particle growth undergoes several sub-stages but in the presence of sufficient air and OH radical the already formed soot particle oxidizes (typically in the later stage of combustion towards the end of injection and mixing-controlled combustion in which the mixture becomes less heterogeneous and the temperature drops to a level favoring oxidation over formation) reducing the end level of soot emission. Soot found in the exhaust is usually much less than the soot formed during combustion within the cylinder. Figure 2-7 illustrates a typical history of in-cylinder soot formation.

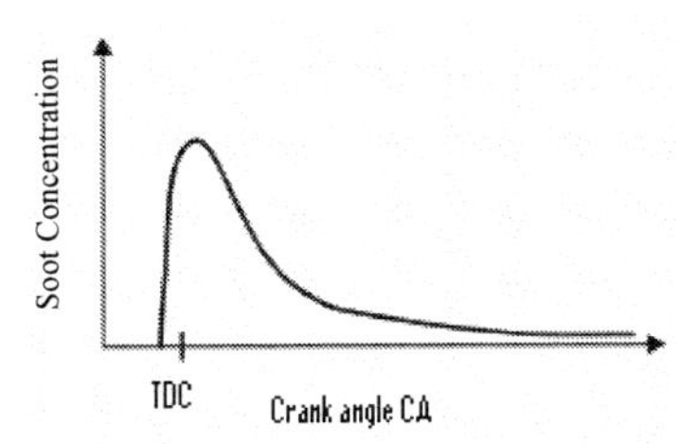

Figure 2-7: Soot concentration as function of CA

A conflicting fact is that both soot formation and soot oxidation are favored at high temperatures but the later also requires sufficient oxygen concentration. It is, therefore, the goal of soot emission control in diesel engines to come up with a combustion strategy in which temperature during soot formation is lowered while temperature during oxidation is increased. In addition, to further reduce soot formation, fuel introduction into the cylinder through the injection system should be done in a way that enhances mixture formation aiming at increasing its homogeneity and reducing the portion of fuel being injected onto burning flame or hot gases.

To better understand soot formation in diesel engine one needs to consider its evolution and the physical and chemical processes it undergoes.

The evolution of liquid-or vapor-phase hydrocarbons to solid soot particles generally involves six different recognizable processes: fuel pyrolysis, nucleation, coalescence, surface growth, agglomeration and oxidation.

A schematic illustration of the sequence of the first five of these processes compromising the soot formation process is depicted in Figure 2-8. The sixth process, however, is soot oxidation in which hydrocarbons, already formed soot particles inclusive, are converted to CO, CO_2 and H_2O at any point in the process. This process as illustrated can proceed in a spatially and temporally separated sequence as encountered in a laminar diffusion flame or may occur simultaneously as is the case in a well-stirred reactor. In practical combustion systems however the sequence of processes can vary between these two extremes as the case of diesel combustion in which combustion takes place in a premixed as well as non-premixed diffusion flames.

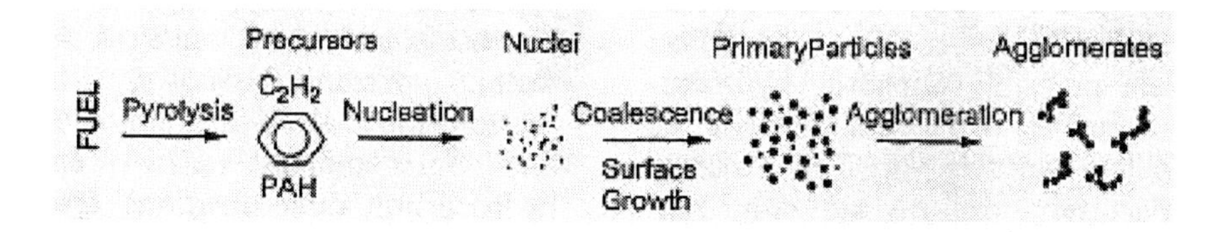

Figure 2-8: Schematic diagram of the steps in the soot formation. From Tree and Svensson [16]

Fuel pyrolysis

Pyrolysis describes the process in which fuel compounds alter their molecular structure under high temperature without significant oxidation although oxygen species may be present. It results in the production of precursors or building blocks for soot and its chemistry is often viewed as playing the dominant chemical role in determining the soot propensity [17]. Soot precursor formation which, as already mentioned the result of fuel pyrolysis, is a competition between the rate of pure fuel pyrolysis and the rate of fuel and precursor oxidation by the hydroxyl radical OH. Both pyrolysis and oxidation rates are temperature-dependent and increase as temperature increases.
As a result premixed flames tend to soot less than diffusion flames because of the present of oxygen which facilitate and accelerate oxidation rate.
Furthermore the incipient of particle formation in diffusion flames depends to a certain degree on the H atom concentration which accelerates pyrolysis when diffused into a fuel-rich zone [17, 18]. Glassman [17] adds that larger molecules increase the radical pool size which in return promote and accelerate pyrolysis. Although the extent of fuel conversion into soot may very well vary from one fuel type to another, a common mechanism of soot formation exists and majority opinion at present is that soot particles form through the pyrolysis product of Polycyclic Aromatic Hydrocarbons [PAH] [19], especially acetylence.

The primary focus is actually on the *formation of the first aromatic ring* from small aliphatics, given that this step is perceived as rate-limiting step in the reaction sequence to larger aromatics. Several possibilities for the formation of the first ring were proposed. Initially the even-carbon-atom pathway involving the addition of acetylene to n-C_4H_3 and n-C_4H_5 was proposed but soon has been dismissed by Miller and Melius [20] who argued that n-C_4H_3 and n-C_4H_5 could not be present in high concentrations because they transform rapidly to their corresponding resonantly stabilized isomers and proposed instead, along with others, an odd-carbon-atom pathway via combination of propargyl radicals.

$C_3H_3 + C_3H_3 \rightarrow$ benzene or phenyl+H (reaction 1)

Frenklach et. al. [21] have been arguing, however, that formation and growth of aromatics is initiated by a propargyl-acetylene reaction and has been recently supported by Moriatry et al. [22] and many others. Thus the proposed reaction is

$C_3H_3 + C_2H_2 \rightarrow$ c-C_5H_5 (reaction 2)

This pathway combines highly stable radical, propargyl, and the most abundant "building block", acetylence. Once formed cyclopentadienyl reacts rapidly to form benzene [23,24,25].
To determine whether inception of soot formation, or more accurately formation of first aromatic ring goes indeed via reaction (1) or (2) a comparison between reaction rates of these reactions has to be considered as follows:

The ratio of reaction rate of reaction (2) to (1) is:

$$\frac{k_2[C_3H_3][C_2H_2]}{k_3[C_3H_3][C_3H_3]} \sim (0.02 - 0.1)\frac{[C_2H_2]}{[C_3H_3]}$$

(where $[C_3H_3]$ cancels out and $k_3 = (1 \text{ to } 5\times10^{12})\text{cm}^3 mol^{-1}s^{-1}$ was used based on values given by [26,27,28]).

The concentration ratio $[C_2H_2]/[C_3H_3]$ is of the order of 10^2-10^4 as reported in many experimental flame studies (*e.g.* [29]-[32]). This yields that reaction (2) is expected to be faster than reaction (1) by a factor of 2-10^3.
This implies indeed that reaction (2) is not only faster enough but should most probably play a dominant role in the formation of the first aromatic ring.

Following the formation of the first aromatic ring, two subsequent processes, namely *aromatic growth* and *aromatic oxidation*, would compete among themselves. The domination of the former determines the progress of soot formation. The mechanism of aromatic oxidation seems to be the oxidation by O_2 [33].
As a matter of fact, these two processes are often viewed as related ones since while the H atom production increases aromatic growth given its role in accelerating the abstraction of hydrogen atom from the reacting hydrocarbon and thus promoting the subsequent step of converting aromatic molecule to a radical leading to aromatic growth, it destroys O_2 leading to a declination in oxidation. In other words, as aromatics growth progresses, it should be accompanied by the depletion of O_2 [19]. This explains why soot inception appears often in the vicinity of the main combustion zone, in an environment rich in H atoms and poor in O_2 molecules. The effect of aromatic oxidation takes place in the form of removal of carbon mass from further growth which otherwise would contribute to the further growth of aromatic.

Nucleation

Following the formation of the first aromatic ring, the transition of the gas-phase species to solid particles takes place in a process known as nucleation. The already formed aromatic ring is thought to add alkyl groups, turning into a PAH structure. Thanks to the presence of acetylene and other vapor-phase soot precursors, PAH grows to a size large enough to develop into a particle nuclei.
This process is restricted to a region near the primary reaction zone where temperatures, radicals and ion concentrations are the highest in both premixed and diffusion flames [34]. According to Glassman [18], the processes of growth to larger to larger aromatic ring structures leading to soot nucleation and growth are similar for all fuels and faster than the formation of first ring.

Surface growth

Once aromatic ring develops into particle nuclei a process of mass addition to the surface of the nucleating particle takes place leading to a surface growth of this particle. This can take place concurrently to nucleation. It is the abundance of acetylene that is believed to be the prime promoter of surface growth [19, 16]; so that the hot reactive surface of the soot particle readily accepts gas-phase hydrocarbons leading to an increase in soot mass. The intensity of this mass increase is thought to be temperature and particle size dependent [35]. Since smaller particles have more reactive radical sites, their surface growth rate is higher. Surface growth continues as the particles move away from the primary reaction zone into colder and less reactive regions. It is believed that the majority of soot mass is added during surface growth. It is worth noting, therefore, that the residence time of the surface growth process has a large influence on the total soot mass or soot volume fraction. Like all processes, surface growth is affected by slowing and declination factors.
These factors were identified by numerical simulations performed by Frenklach and Wang and Frenklach [36, 37] who concluded that a decrease in H atom concentration or a degree of H super equilibrium in the gas phase and a decrease in the number of active sites on the soot particle surface capable of sustaining growth due to surface defect leading to what is termed *surface aging* which is the declination of surface growth.

Coalescence and agglomeration

Once soot particles are formed, they collide with each other forming larger particles. Coalescence occurs when particles collide and coalesce. In this process particles are assumed to be spherical and the resulting larger particle following coalescence is also spherical. This process seems to be pressure-dependent because its occurrence is related to the ratio of the mean free path to the particle radius known otherwise as the Knudsen number. Agglomeration on the other hand occurs when individual or primary particles stick to each other forming chain-like structures. Agglomeration also continues after combustion ends and form chain-like structure larger in size, typically in the range of 100nm-2μm [38].

Oxidation

Oxidation is the process of converting carbon or hydrocarbon to combustion products. Here one needs to distinguish between oxidation of aromatic as already mentioned and oxidation of soot particle. The later can take place at anytime during a soot formation process from pyrolysis through agglomeration. Once carbon has been partially oxidized to CO, the carbon will no longer evolve to a soot particle even if it enters a fuel-rich oxygen-poor zone. Soot oxidation is mixture state and temperature dependent. It occurs at temperatures higher than 1300 K while the graphite-like structure of soot is thought to render soot highly resistive to oxidation. In general, oxidation of small particles is thought of as a two-stage process. In the first stage, oxygen is chemically attached to the surface hence, oxygen absorption and in the second desorption of oxygen with the attached fuel component from the surface as a product takes place. Unlike in the oxidation of aromatics, OH seems to dominate soot oxidation under conditions related to diesel engines. Presence of oxygen promotes, however oxidation by increasing the OH radical pool.

Now one needs to consider the diesel jet spray structure which has already been demonstrated in the previous sub-section in order to attain an insight on the locations and evolution of soot formation within the diesel spray. Figure 2-9 illustrates schematically the overall spray structure and the locations where soot formation is doomed to take place. As depicted in the illustration, a diffusion flame surrounds the entire region downstream of the liquid penetration and extends upstream to a fixed distance from the nozzle tip. This fixed distance is termed the *Lift-Off Length* [LOL] which according to some studies seems to have great influence on soot formation as would be later discussed.

According to measurements performed by Dec [39] and later by Dec and Westbrook [40] and Dec and Tree [41] the general evolution of soot formation is generally perceived as a multi-stage process. The early soot seems to be formed in the fuel-rich premixed reaction, at the beginning of what is usually considered as the premixed combustion spike of heat release. As the fuel passes this rich premixed reaction, it yields hydrocarbon fragments that become the building blocks of soot particles inside the diffusion flame sheath. As they travel down the spray plume they agglomerate and react to form larger particulates. However, as it is transported toward the boundary of the diffusion flame, a heavy presence of OH radicals seem to lead to particulate oxidation while high heat release rates persist, thus satisfying both conditions for the occurrence of soot oxidation, namely, sufficient OH and oxygen. This hypothesis is supported by the Laser-Induced Incandescence [LII] and Planer Laser-Induced Fluorescence [PLIF] OH measurements performed by Dec and Tree [41] who showed soot within the jet being perfectly enveloped by OH and no soot was visible outside this envelop. Diesel soot emissions appear, therefore, to be the result of the quenching of this final phase of oxidation. Furthermore, Siebers and Higgins [43] have concluded that when soot is not produced in the rich premixed zone, it is also not produced in significant quantities later as the right products react in the diffusion flame.

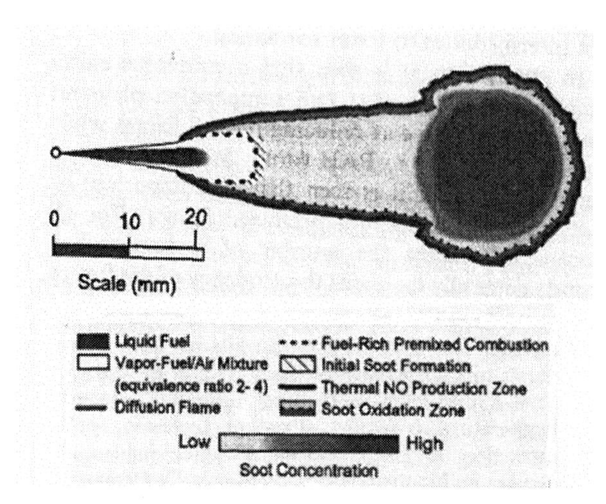

<u>Figure 2-9:</u>
Conceptual picture of a reacting diesel spray during the quasi-steady portion of combsution from Dec [39]

Given that soot is initially formed in the rich premixed combustion as mentioned above, the amount of fuel mass entrained into air and thus the amount of premixing of fuel and air would have great influence on the amount of soot formed. Siebers and Higgins [43,44] have shown that it is the LOL, not the liquid length that controls the level of ambient gas premixing prior to combustion within the rich premixed zone. Siebers and Higgins also showed that as LOL increases, relative quantities of soot decrease until a point is reached where the percentage of stoichiometric air in the range of 40-50%. At this point, soot emission becomes undetectable. They have concluded, therefore, that the parameters that control the LOL and thus, the amount of entrained oxygen, are those that control soot formation. Such parameters are physical parameters such as in-cylinder pressure and temperature and engine parameters, namely injection pressure, as well as fuel parameters such as Cetane Number whose influence would be discussed in the subsequent sub-sections.

2.3 Parameters Influencing Diesel Combustion and Soot Formation

Many parameters affect directly or indirectly diesel combustion and subsequently soot formation. Some are related to engine parameters, mainly, injection pressure and timing, others to physical parameters such as in-cylinder gas temperature and pressure while several others relate to the parameters of the injector itself, mainly injector nozzle geometry and injector needle activation mechanism. In the following sub-sections we shall shed the light on these key parameters.

2.3.1 Effects of Engine Parameters

Injection pressure has a paramount effect on diesel combustion given its influence on spray development and mixing formation. As fuel injection systems have undergone tremendous developments and improvements modern fuel injection systems can pressurize fuel to pressures exceeding 2000 bars. The advantages of higher injection pressures are diversified. Increasing injection pressure increases the fuel jet momentum and velocity. This leads to higher fuel entry velocity and faster jet penetration. If considering equation 2-1 increasing the jet velocity increases its inertia relative to the compressed in-cylinder gas. As a result the jet break-up and its atomization rate increase (bigger Weber-numbers) significantly. Subsequently following stronger momentum exchange between the incoming fuel jet and the surrounding compressed gas due to higher injection pressure the fuel breaks into smaller droplet size resulting in faster evaporation and air-to-fuel mixing.
Several research works have been focused on investigating the effect of injection pressure on fuel jet penetration. Many have even arrived through direct visualization of in-cylinder spray development to the formulation of empirical equations relating spray penetration length to injection pressure or more accurately to the pressure drop over the injector nozzle. One remarkable formulation is that of Hiroyasu [45]. He gives the following relation for the spray penetration length S:

$$S = 0.39\sqrt{\left(\frac{2(\Delta p / pascals)}{(\rho_f / Kg / m^3)}\right)}\,(t/s) \qquad \text{for } 0 < t < t_{br} \qquad (2\text{-}7)$$

$$S = 2.95\left(\frac{(\Delta p / pascals)}{(\rho_g / Kg / m^3)}\right)^{0.25}\left((d_0 / m)(t/s)\right)^{0.5} \qquad \text{for } t > t_{br} \qquad (2\text{-}8)$$

Where:

$$t_{br} = 28.65(\rho_f d_0)(\rho_g \Delta p)^{-0.5}$$

Ofner [46] has also performed measurements of in-cylinder spray development and reported similar trends but with different values for the empirical coefficients.
He too argued that the jet penetration velocity increases with increasing pressure injection and that fuel penetration length increases proportional to the square root of the pressure drop over the injector nozzle.

Stegemann [2] and Kamimoto et al. [47], on the other hand argued that while the fuel jet velocity indeed increases with increasing injection pressure, the liquid fuel penetration length is almost independent of injection pressure. This is because while the jet momentum increases with pressure which presumably should lead to higher penetration length, the evaporation rate of the atomized fuel increases as well leading to an increase of the penetration length of the gas phase with no change in the length of the liquid fuel. In other words the expected increase in liquid fuel penetration length with increasing injection pressure is balanced out with the stronger evaporation. Nevertheless the maximum penetration length of the liquid phase is reached earlier (faster) with increasing injection pressure [48]. It seems, however, that the dependency of fuel jet penetration length on injection pressure is not monotonic. At certain injection pressures, further increase in injection pressure leads to a little difference in spray penetration. Morgan et. al. [49] have observed a weak dependency of spray penetration on injection pressure for pressures ranging from 130 to 160 bar. This can also be seen from equation 2-8 where the dependency of spray penetration on injection pressure has a weight of one forth of a power. This indicates that the effect of a change in injection pressure on jet penetration would be reduced at higher injection pressures. As a result; its doomed that the key advantage of higher injection pressures lies in its effect on jet velocity and in improved atomization which subsequently leads to faster droplet evaporation and better air-fuel mixing.

While change in injection pressure has significant effect on the fuel jet velocity, its penetration and atomization rate it seems to have no significant effect on the fuel spray cone angle which always has been considered as a measure for air-fuel entrainment and thus air-fuel mixing. Ofner [46] has also reported that increase in injection pressure seems to have no recognizable effect on spray cone angle. Such behavior has been explained that while the jet velocity increases with increasing injection pressure the droplet size decreases as mentioned above leading to an increase in evaporation rate. Thus the actual spray cone angle remains unchanged.

Pischinger et al. [50] have also reported increase in ignition delay with increase in injection pressure. This seems to be attributed to the decrease in local mixture temperature due to accelerated evaporation.

Despite the paramount effect of injection pressure on diesel combustion only few attempts were made to investigate the influence of injection pressure and timing on soot formation [51]. The pioneer work of Picket and Siebers [52] utilizing Laser-Induce Incandescence imaging has given a valuable insight of the effect of injection pressure on in-cylinder soot formation. They have shown that soot level in the fuel jet decreases substantially with increasing injection pressure and the peak soot decreases linearly with increasing injection velocity (i.e., the square root of the nozzle orifice pressure drop). This is mainly because lift-off length increases linearly with increasing jet velocity [53]. This in return increases the total air entrained into the fuel jet upstream of the lift-off length relative to the amount of fuel injected leading to a decrease in the local equivalence ratio and thus decrease in the local mixture temperature. Another factor contributing to the decrease in soot level with increasing injection pressure is the effect of residence time in the reacting region on soot formation. With increasing injection pressure, hence increasing jet velocity, the time for a fluid element to move through the soot-forming region of the fuel jet decreases.

There is, therefore, less time for soot formation before the oxidation-dominated region is reached. Furthermore, given the increase in lift-off length with increasing injection pressure, the length of the combustion region is shortened, resulting in less time for soot formation and growth. This also implies that combustion duration is shorter with increasing injection pressure leading to an earlier soot burnout in the expansion stroke when temperature and soot oxidation rates are relatively higher.
Other works have also shown that with increasing injection pressure (up to 100 MPa) soot was located further downstream of the spray [54,55,56]. Crua et. al. [51] who have also performed LII imaging of diesel soot have reported an earlier appearance, i.e., formation of soot for higher injection pressure (140 and 160 MPa). They too have reported lower peak soot with increasing injection pressure. Overall it was found that higher injection pressure produced smaller quantities of soot. This falls in agreement with other observations that higher injection pressure promotes the production of smaller fuel droplets [57] which in return results in reduced fuel vapor concentration leading to the formation of smaller soot particles. These smaller soot particles are then oxidized more rapidly. Also the higher jet momentum associated with increase in injection pressure as already mentioned promotes faster and improved momentum exchange between the liquid spray and the surrounding cylinder gas. This, in return leads to a better air-fuel mixing and to an increase in local oxygen content, i.e., decrease in local equivalence ratio resulting in lesser soot formation and more soot burn-out. Herrmann et. al. [58] have also reported similar results where an increase in injection pressure has produced less soot level. They argued that the increase of jet momentum with increasing injection pressure enhances the turbulent mixing resulting in accelerated soot oxidation.

The effect of **injection timing** on combustion and soot formation is directly related to the effect of cylinder temperature and pressure on combustion and soot formation which would be discussed in later sub section but generally speaking by advancing injection, cylinder gas temperature and pressure at the time of injection are lower leading simultaneously to a longer ignition delay and to a slower atomization rate. This allows longer time for fuel-air mixing but results also in higher combustion temperature and pressure. Soot formation could as a result very well increase if a greater fraction of the fuel is burned at higher temperature but because the end of injection comes earlier, the temperature is higher facilitating faster soot burnout [16]. It is doomed that the effect of injection timing on soot oxidation and burn out is more critical. If injection timing is retarded the cylinder gas temperature and pressure at the time of injection are higher and although it could produce more soot due to shorter ignition delay, lower local equivalence ratio and higher temperature, soot oxidation could be further accelerated due to higher temperature and better mixing rate associated with improved turbulent flow during the expansion stroke. This leads to lower final soot level.

2.3.2 Effects of Injector Parameters

Key injector parameters, injector nozzle and needle activation mechanism type in particular, have a decisive effect on diesel combustion and emission. They directly affect spray characteristics which in return determine mixture formation and combustion development.

The Nozzle as the determining interface between injection system and combustion chamber encompasses several critical factors namely, nozzle design distinguished by sac hole and seat hole and nozzle geometry distinguished by key factors such as orifice diameter, hole conicity referred to as K-factor (defined by inlet and outlet diameters according to $K = ((D_{inlet}/\mu\text{m}) - (D_{outlet}/\mu m))/10$), rounding radius at the inlet orifice Ra, the inlet edge and nozzle's hole number.
Nozzle design shown in figure 2-10 is known to affect soot and Hydrocarbon [HC] emissions. A sac hole nozzle exhibits symmetric sprays compared to asymmetric spray as the case of seat hole which yields the advantage of increasing air entrainment resulting in lower soot emission. It yields, however, higher unburned Hydrocarbon as the fuel mass trapped in the sack volume enters the combustion chamber during the closing of the nozzle needle at very low pressures. This fuel mass is poorly atomized and because of its low momentum doesn't penetrate the combustion chamber and remains close to the nozzle tip. As a result it burns incompletely and contributes further to the unburned HC.

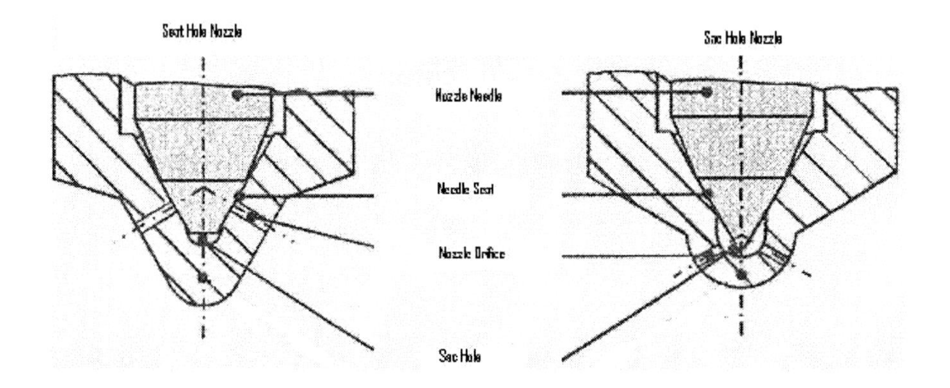

Figure 2-10: Illustration of various nozzle design. From Badock [65]

Despite the disadvantage of higher HC emission sac hole remains favorable over the seat hole nozzle because of its improved soot emission characteristics. They have undergone several improvements to improve their HC emissions. The focus of such improvement lies in the reduction of the sac volume. Three designs are already available as serial production shown in figure 2-11., Mini-sac, macro sac and Valve Covered Orifice [VCO]. The later has the smallest sac volume but to further decrease HC emission the nozzle seat completely covers the nozzle orifice at the end of the needle closing. Potz et. al. [59] have shown that the VCO configuration yields on one hand the lowest HC emission but worse spray symmetry on the other because of the typical asymmetric velocity profile at the outlet. Generally speaking, the smaller the volume sac is the lower HC emissions are but the deteriorating spray symmetry is.

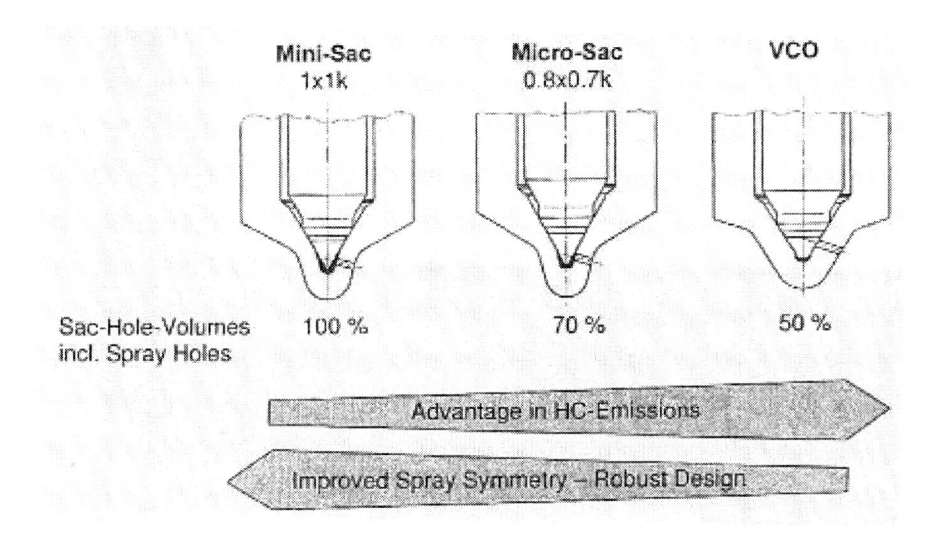

Figure 2-11: Illustration of performance of various sac designs. From Potz et. al. [59]

Nozzle geometry on the other hand exhibits more complex effects. The various geometric parameters affect directly the internal flow within the nozzle which as a result determines the spray pattern and characteristics. The core of the geometric effect lies in the tendency to induce cavitating flow. Two types of cavitation can emerge within the flow, *dynamically-induced* cavitating string vortices and geometrically-induced hole cavitation. The former are formed in the sac volume and generated by the needle motion which enter the nozzle holes while the later occurrs at the hole entrance and is generated as a result of a turning the flow experiences as it enters the nozzle hole. The later is of a paramount effect. At the nozzle inlet the flow experiences turning because of the inlet angle and edge rounding radius leading to acceleration within the flow and as a result a pressure drop. When pressure drops below the fluid vapor pressure a cavitation zone initiates and forms a vena contracta inside the orifice. Nurick [60] has presented a simplified diagram of this phenomenon and is shown in figure 2-12. Such contraction in the inlet section reduces the physical area through which the liquid flows resulting in an increase in flow velocity. Chaves and Obermeir [61] have shown through their flow measurements that velocities at the nozzle exit are higher than those deduced from velocity discharge measurements which can be attributed to the reduction in the orifice cross section.
The emergence of two-phase flow because of the onset of cavitation can influence fuel atomization. Bergweg [62] has argued that the collapse of cavitation bubbles in the nozzle holes is expected to enhance fuel atomization through generation of smaller droplets that vaporize faster enhancing fuel-air mixing and reducing ignition delay. Arcoumanis and Whitelaw [63] have described this effect as the low-temperature vaporization of the more volatile component of the multi-component diesel fuel. In this context, geometric dimensions can either promote or suppress the onset of cavitation.

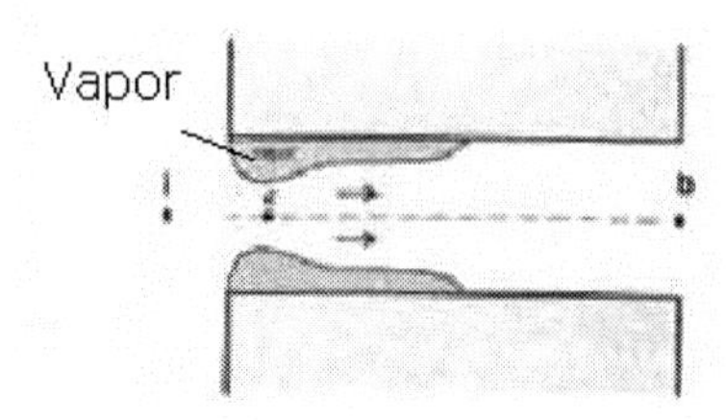

Figure 2-12: Schematic of cavitation zone forming vena Contracta.

Arcoumanis et al [64] and Badock et al [65] who have investigated cavitation phenomenon in real-size diesel injection nozzles have shown that the hole inlet edge, its radius and conicity have great effect on cavitation. In order to evaluate how such geometric dimension can affect the onset of cavitaion, a measure for the onset of cavitation has been formulated and is termed cavitation number [KN] defined as:

$$KN = \frac{P_i - P_b}{P_i - P_{vap_i}} \tag{2-9}$$

Where P_i is the injection pressure, P_b is the back pressure and P_{vap} is the fuel vapor pressure.

The onset of cavitation is then determined by comparing KN to a critical value termed KN_{crit}. Macia et al [66] have shown through CFD calculations that cavitaion occurs when $KN^{1/2} > KN_{crit}^{1/2}$ while KN_{crit} was not only nozzle geometry-dependent but injection pressure-dependent as well. A general empirical correlation for KN_{crit} has been given as:

$$KN_{crit} = \mathrm{Re}^{-0.01} \cdot Ra^{0.178} AR^{0.022} D_{out}^{-0.496} \tag{2-10}$$

Where Re is the Reynolds number which encompasses injection pressure by recalling the mass continuity and Bernoulli equations, $U = \frac{\dot{m}}{A\rho}$ and $P_0 = P_1 + \frac{\rho}{2} \cdot u_1^2 = P_2 + \frac{\rho}{2} \cdot u_2^2$ respectively, Ra is the inlet rounding radius, AR is the area reduction of the nozzle because of nozzle conicity given as $\left(D_i^2 - D_{out}^2\right) / D_i^2 \times 100$, and D_{out} is the diameter at the outlet orifice of the nozzle.
It was shown that for certain conicity, namely larger conicity, and certain inlet rounding, cavitation can be suppressed completely. The tendency is, therefore, to reduce outlet orifice diameter and increase nozzle hole number.

Furthermore, Payri et al [67] have shown that sprays of conical nozzle penetrate on average 10 % higher than that of cylindrical nozzle. This is attributed to the velocity increase caused by the nozzle conicity. It decreases on the other hand spray cone angle and thus decreases air entrainment. Actually Chavez et al [68] have observed increase of spray cone angle as a result of cavitation.

They referred to a clear zone where the spray cone angle increases and afterwards a stabilization zone where cavitation is assumed fully developed.

Number of holes of nozzle has also significant effect. It is, however, coupled with hole diameter. The analysis of the effect of number of holes of a given nozzle on combustion and emission is done while maintaining the overall total injected fuel mass unchanged. Meaning to say reducing hole diameter by increasing number of holes and vice versa so that the overall total chemical energy added to the system remains the same.
Eisen [69] has performed interesting investigation of the effect of number of holes through visualization of the in-cylinder events and his results have shown that for smaller number of holes the injected fuel mass per hole increases resulting in larger spray momentum. This allows in return faster spray penetration and stronger momentum change with the surrounding air yielding faster evaporation rate on one hand but a stronger wall impingement on the other. Larger number, however, allows a more uniform distribution of fuel within the combustion chamber and higher total surface area that comes with contact with the surrounding air enhancing also air entrainment which in return increases evaporation rate and reduces evaporation time. Nevertheless, no clear effect on ignition delay could have been confirmed. It was observed, however, that larger number of holes tends to yield almost no or minor cycle-to-cycle variations in the ignition behavior.
The effect of number of holes on soot formation is twofold. On one hand the weaker momentum associated with larger number of holes reduces the lift-off length which in return increases the tendency to soot, while on the other hand the faster evaporation rate associated with it enhances air entrainment and decreases as of a result the propensity to soot.

The needle activation mechanism of an injector also affects spray and combustion characteristics. The two common types are *solenoid* and *Piezo-electric* valves. In many aspects the piezo is thought to be superior to solenoid valve. It is almost 3 times faster in lifting and closing a needle yielding as a result lower throttling losses and pressure drop at the hole inlet increasing the injected fuel mass momentum which in return penetrates faster. The faster needle activation time of a piezo reduced also fluctuations in injected mass reducing, therefore, cyclic variations.
The stronger spray momentum associated with piezo injector enhances air entrainment because of the stronger momentum change between the injected fuel and surrounding air improving thus air-fuel mixing and increases the fraction of pre-mixed combustion. The later increases however cylinder pressure and temperature rise rate and it was observed experimentally that solenoid valves tend to have a weaker pressure rise indicating higher rate of diffusion combustion. Nevertheless, both injector types seem to exhibit similar maximal spray penetration length and almost no difference in the start of combustion. Piezo injector, however, promotes faster combustion which could be desirable if the effect of residence time on pollutant formation is to be considered.
Also Ofner [46] has observed optically that the solenoid valve type of an injector tend to yield a larger spray cone angle at the beginning of the fuel injection. This could be attributed to cavitating flow generated from the higher throttling losses and pressure drop at the hole inlet as compared to piezo injector.

Piezo injectors are as of result more favorable in particular for injection strategies where injection of a smaller amount of fuel mass is used such as pilot and split injections.

2.3.3 Effects of Injection Strategy

Single injection strategy is long proved incapable of meeting the more and more stringent emission regulations. The high steep pressure and temperature rise of the premixed combustion of a single injection strategy yields high combustion noise and introduce a complex trade-off between the various emitted pollutants. Varying injection timing alone does not bring simultaneous reduction in key pollutants such as soot and NOx. On the contrary it only benefits one element at the expenses of the other. Figure 2-13 shows the trade-off associated with injection timing in terms of soot and NOx emissions.

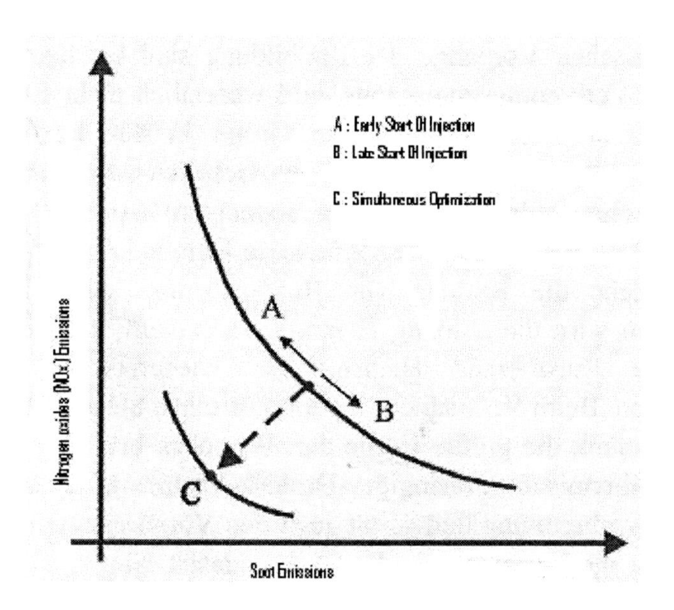

Figure 2-13: Schematic of the effect of injection timing on NOx and soot emissions

A promising approach to come along with such complex trade-off is to split the fuel combustion process into two (or more) parts with the intention to generate two (or more) separate heat release peaks triggered by two (or more) injections as depicted in figure 2-14

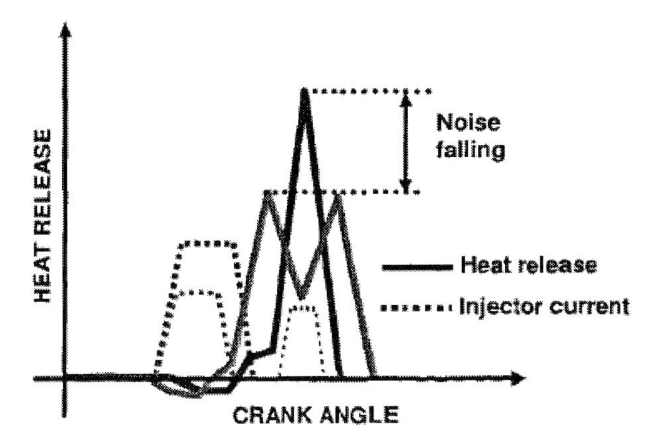

Fig 2-14: Schematic of split injection effect on heat release.

Split injection generates, therefore, a lower maximum amplitude of the split heat release than the peak of heat releases realized with a single injection. This in return leads to a lower combustion noise. Depending on the auto-ignition delay of the second injection, the split in heat release may correspond to two different combustion processes [70]:

1. two premixed type combustion phases or,
2. one premixed type combustion followed by a diffusion combustion of the second injection.

The split is realized in several patterns depending on the desired combustion development and heat release rate for a given combustion system or strategy. Thanks to the advancements in needle lift control of the injector the injection split can be realized in what is termed **Multiple Injection** [MI]. Multiple injection strategies already deployed in commercial diesel engines such of those installed in modern buses to satisfy the European emission regulations of the so-called Euro IV(4) and Euro V(5) emission regulation consists of Pilot Injection, Main Injection and Post Injection.

Pilot injection is primarily used to reduce noise which is caused by the steep pressure gradients (dp/dt)due to the rapid combustion of the so-called premixed combustion [71]. Its direct effect, however, can vary depending on the combustion system and its associated cylinder temperature and oxygen concentration.
In a typical high compression ratio engine, the direct effect of a pilot injection is a decrease in the ignition delay of the main injection resulting in a less rapid combustion with less steep pressure gradient which in return lowers maximum cylinder temperature yielding significant reduction in NOx emissions [72].

As pertains to soot emissions the timing of the pilot injection or more accurately the time interval between it and the subsequent main injection plays a major role.

Vanegas et. al. [72] have shown that for fixed timing of the main injection, the time interval between main and pilot injection affects directly the ignition delay of the main injection. An intermediate time interval (an injection timing of the pilot set to a time between advance and late timings) caused the shortest ignition delay while an early pilot injection yielded the lowest smoke emissions (mainly soot) level. Overall smoke emissions for a split injection strategies were, however, significantly higher than those of a single injection strategy. Helmantel and Golovitchev [73] added, however, that advancing pilot injection to certain degree can increase soot emissions but it declines sharply afterward and further advancement of the pilot injection stabilizes it. Furthermore, advancing pilot injection decreases the portion which burns before the main combustion. A very early pilot injection can result in no heat released before the main combustion. With advanced timing of the pilot injection, the fuel injected will be premixed before igniting and its contribution to the soot formation will be limited or near zero. The actual timing of the pilot injection is usually determined, therefore, in accordance with best trade-off between soot and NOx emissions as well as thermal efficiency.

In combustion systems with low compression ratio and relatively high EGR rates, pilot injection can, on the other hand, affect combustion in a different way. In such system the in-cylinder temperature at the time of injection is low resulting in a significantly longer ignition delay and the auto-ignition delay exhibits strong sensitivity to the cooling effect that is induced by the fuel evaporation [70]. The time interval between pilot and main injection can be used in such combustion system in a manner to exploit the cooling effect.
The temperature drop brought about by fuel vaporization may slow down fuel oxidation of the first injection delaying its start of combustion. The ignition delay of the second injection is, however, shorter than the ignition delay of the first injection.

The investigations of Chang Lee et. al. [74] have shown that split injection has also a direct effect on the spray and atomization characteristics. It has been observed that the spray tips of both first and second injection event in the case of split injection are flatter than that of single injection. This has been attributed to the lower jet velocities associated with the split injection as the momentum of each injection event of the split injection is fairly lower than that of the single injection. Furthermore, the spray width in the split injection case is increased in comparison to the single injection case. As per the spray tip penetration, it was observed that the second injection has a longer penetration length than that of the first injection. This is mainly because of two reasons. First the injected fuel mass in the split injection case is split into two injection events. The momentum of each is according to the actual mass split ratio but overall lower than the momentum of a single injection. It's accustomed that the fuel mass injected in the first injection event is much smaller than that of the second injection as a result the actual jet momentum of the second injection is larger than the first bringing about larger penetration length. Second, the entrained flow of ambient gas directed downstream created by the first injection causes a reduction in the drag on the subsequent injected fuel spray. This also reduces the relative velocity of the second injection which in return generates droplets with larger SMD. However following atomization of the second injection the overall droplet size is reduced again. This increase in droplet size even temporarily accompanied with longer residence time associated with split injection can lead to an increase in the rate of soot formation.

Post injection, on the other hand, is primarily used as a third injection event of a multiple injection strategy. It is used as a measure to counter the increased production of soot brought about by the split injection and thus recover the trade-off between soot and NOx emission levels. There are several theories explaining the effect of post injection on soot reduction. These can be grouped in three categories:

1. post injection introduces extra energy for mixing which in return improves the oxidation process and reduces the final soot level [75,76,77,78,79]
2. the high temperature in the chamber prompted by the post injection is responsible for the improved oxidation and, consequently, for a reduction in the final soot emissions [80,81]
3. The reduction in the final soot emissions level decreases because the main injection produces less soot and the post injection does not produce significant additional soot [82,83]

Despite the various theories, the timing of post injection remains the prime dominant factor since as the injection timing is modified, the thermodynamical and flow conditions during the combustion process also change. Several investigations have been conducted for this purpose. It has been shown that if post injection is retarded to an intermediate timing soot emissions level rises while it again declines if timing is further delayed [72,73,84]. Vanegas et. al [72] have concluded that smoke emissions can be reduced compared to split injection strategy when applying early post injection. They argued that if post injection is done during combustion, the soot particles are burned and the pressure of the post injection possesses sufficient kinetic energy to achieve an effective mixture formation. Helmantel and Golovitchev [73] have, however, attributed the pattern of the effect of post injection timing on the reduction of soot emission to the effect of turbulence and its effect on enhancing soot oxidation. They have also argued that a possible explanation for the rise in soot emission for the case of intermediate timing of post injection could be that the spray of the post injection enters a low oxygen region which in return yields increase in soot formation rather than an increase of soot oxidation.
Mendez and Thirouard [70] have also performed similar investigation but they concluded that in order to reduce smoke emission the post injection should occur near the middle of the diffusive flame period. If post injection timing is retarded too weak temperature and not enough time is available to ensure soot post oxidation.
Finally, Arregle et. al. [84] have conducted investigation to confirm that theory 3 mentioned above is the likely applicable one to explain soot reduction when using far post injection. They have indeed observed no significant soot formed during the far (late) post injection and even concluded that soot is unable to be formed under a threshold temperature of 700 K and since the fuel mass injected in the main injection was reduced in the amount equal to that of the post injection, it is indeed true that in such circumstances theory 3 is valid. However, it is necessarily to note (as also the author did in this work) that theory 3 could very well be totally invalid and even irrelevant if the fuel mass injected in the main injection remains unchanged which is a necessary measure to maintain engine torque equal to that of the case of split injection (double injection event with no post injection). In this case theories 1 and 2 could be the only applicable explanations.

2.3.4 Effect of Compression Temperature and Pressure

The effect of compression temperature and pressure is twofold. They affect mixture formation on one hand and combustion and soot formation chemistry on the other.

Temperature effect on mixture formation is sensible from the perspective of temperature effect on the ignition reaction, hence ignition delay, as well as on fuel evaporation rate. As has been already demonstrated increase in temperature results in decrease in ignition delay time which in return shortens the time available for air entrainment leading to fuel richer mixture which consequently leads to increase in soot formation. Furthermore, increase in temperature permits higher rates of heat transferred to the liquid fuel spray resulting in accelerated evaporation rates which in return shortens the liquid fuel penetration length and as a result decreases air entrainment.
Increase in temperature on the other hand decreases air density and thus air resistance in the form of drag forces acting on the penetrating fuel jet allowing faster spray propagation which could enhance air entrainment depending on the available residence time.
Pickett and Siebers [52] have shown that increase in temperature leads to a decrease in lift-off length and thus soot processes are shifted to regions closer to the nozzle orifice where local equivalence ratios are higher resulting in increase in soot formation. Furthermore, increase in soot formation because of increase in temperature is also due to the effect of temperature on soot chemistry. It is well known that an increase in temperature results in increased rates of both soot formation and soot oxidation. However, unlike the case of premixed flames, in non-premixed flames an increase in temperature causes an increase in soot due to faster soot formation and lack of significant oxidation chemistry in rich, soot-forming regions [85].

Pressure effect on the other hand which often referred to in terms of gas density is somewhat similar to that of temperature but of different magnitude. Increase in pressure causes increase in gas density resulting in higher air drag resistance. For injection pressures typical of diesel engines this leads on one hand to slower spray propagation while to stronger spray break-up on the other. For a fixed gas temperature air drag effect seems to overtake spray break up effect. As pressure increases the lift-off length decreases and despite possible enhanced air entrainment, soot level increases due to shifting of soot processes closer to the nozzle orifice where mixture is fuel-rich prompting higher soot formation rates. Pickett and Siebers [52] have reported non-linear effect of pressure (or density) on soot formation while Mueller and Wittig [86] have argued that soot formation can scale with the square of the fuel partial pressure.
Pressure has also direct influence on ignition delay and soot chemical kinetics. Like temperature, increase in pressure shortens ignition delay but with smaller magnitude as compared to temperature and increases soot formation reaction rates. In particular, the rates of reactions responsible for the forming of soot precursors such as PAH [mainly C_2H_2] are pressure sensitive and increase as pressure increases [87].
Increase in air pressure for a fixed amount of injected fuel on the other hand can enhance soot post oxidation depending on gas temperature given the higher concentration of oxygen.

Overall the final soot level will be the result of the balance of soot formation and soot oxidation as both are favored by high temperature and high pressure but the later requires the presence of sufficient oxygen concentration.

2.4 Combustion in Gas-To-Liquid Fueled DI Diesel Engine

In order for an advanced diesel combustion system to meet the increasingly stringent emission regulations, such as the European commission's new standard which for example has set starting effectively 2008, 2 g/KWh and 20 mg/KWh as limit emission levels for NOx and PM respectively, known otherwise as EURO V(5) emission standard (in addition the EU shall demand further reductions in exhaust emissions by 2012, 2013), utilization of alternative fuels, other than conventional diesel, is becoming as a result inevitable. Such alternative fuel should, however, bear the potential to simultaneously reduce NOx and PM emissions. Fuels with zero or near-zero sulfur and low carbon content seem to be the best candidates for meeting such requirements.

Gas-To-Liquid [GTL] fuels are emerging as the best alternative fuels from all aspects. Its properties, along with suitable combustion strategy, allows relaxing the PM-NOx emission trade-off and yields simultaneous reduction in their emission levels [88,89,90,91]. The utilization of GTL fuel requires minor engine modifications in comparison to other alternative fuels and offers remarkable improvements in engine maintenance and exhaust after-treatment devices wear [88]. Thanks to the tremendous recent developments in GTL production techniques, costs of GTL fuel, mainly when using natural gas as the gas source, are very competitive in comparison to that of conventional diesel [92].

GTL is technically referring to Fischer-Tropsch [F-T] diesel Fuel. F-T refers to Franz Fischer and Hans Tropsch, who in the 1920's developed the process, which bears their names, through which hydrocarbons can be produced from synthesis gas (Mixture of CO and Hydrogen). The production of diesel fuels using F-T process can be subdivided into three major "steps" as depicted in figure 2-15.

- Formation of synthesis gas (syngas)
- Fischer-Tropsch catalysis, and
- Post-processing or Refining

Syngas can be formed from any carbonaceous material such as natural gas, coal or biomass while natural gas, due to its properties and availability (readily available in many continents worldwide), is the most attractive material and commercial production of GTL uses natural gas, (GTL fuel used in this work is the type of GTL fuel produced from natural gas). The syngas production is the most expensive part of the entire production process of GTL diesel.

It has been estimated that 70% of the total capital and operating cost is spent in syngas production. There are several possible routes for the formation of syngas from natural gas. Discussing the details and differences between these processes are beyond the scope of this work.

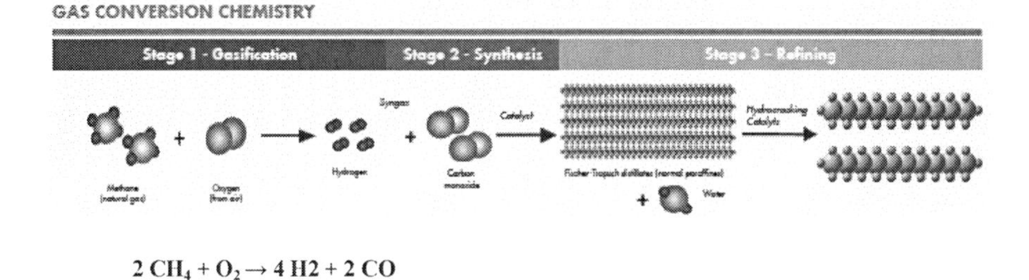

Figure 2-15: GTL/F-T production stages. From Ralphe et. al. [115]

Once syngas is produced, it goes through the F-T process for its conversion into hydrocarbons. This is mainly a catalytic chemical process and the quality of the catalyst used actually determines the quality of the produced fuel. In general, F-T can be divided into two subcategories, low and high temperature. The high temperature F-T processes (300-350°) use an iron-based catalyst and are often used for the formation of gasoline-type hydrocarbons [88]. The low-temperature F-T processes (200-240°), on the other hand, uses an iron or cobalt-based catalyst. Unlike the high-temperature F-T processes, the selectivity for aromatic compounds in the low-temperature F-T processes is low which results in products mixture of relatively pure linear alkanes. Low-temperature F-T processes is, therefore, more suitable for diesel-type hydrocarbons. The end product of the F-T processes lacks, however, the properties enabling them to be usable diesel components mainly due to the high boiling point of the produced alkanes. Further refining through post processing is essentially required to produce usable diesel fuel. In most cases refining relies on hydrocracking. Here again the quality of hydrocracking has a tremendous impact on the end fuel quality and together with syngas production they influence the commercialization of the produced F-T fuel.

Several oil companies have been engaged in the production of GTL fuels. The most successful of which are those who managed to develop high quality and commercially viable syngas formation and refining processes (SHELL for example has a patent on more than 3000 techniques). Some of such attempts are still within lab-scale to pilot-size production plants. Figure 2-16 shows the status of the various GTL production projects per company.

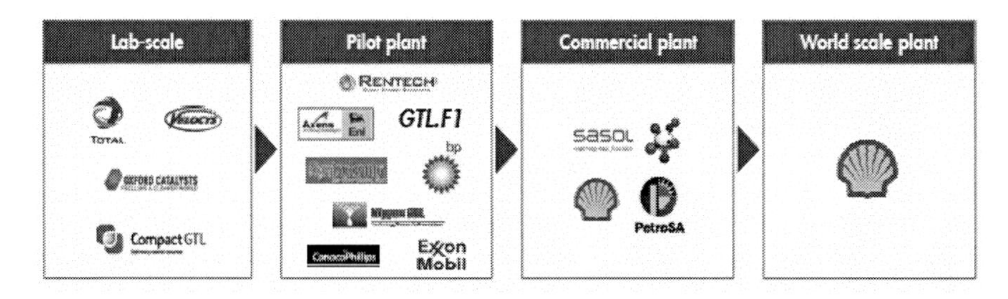

Figure 2-16: Project status of GTL fuel-producing companies

2.4.1 GTL Diesel Fuel Properties

Actual properties of GTL diesel fuel can vary from one manufacturer to another. This is mainly because properties can be controlled by changing certain variables in the synthesis process, including the temperature, catalyst selection, and syngas production method. Overall, all GTL fuels share common properties as follows:

Sulfur content is near zero. The sulfur content in diesel fuel required, by certain regulations, to be zero or near zero to reduce smoke and PM emissions. Sulfur compounds are a potent poison for the catalyst employed in F-T synthesis. It is almost wholly removed during syngas production. GTL fuels contains, by virtue of production, almost no sulfur.

Density of GTL fuel is relatively lower than conventional diesel. This is mainly because GTL fuel is a mixture of normal and iso paraffins which have the lowest density of all hydrocarbons. The density of neat GTL is normally lower than that of diesel by about 7%.

Viscosity of GTL exhibits unclear trends. For some manufacturers it tends to be slightly higher, about 3% higher than diesel [90] while for others (as the case of the GTL used in this work) it tends to be slightly lower, about 8% lower than that of diesel. The author believes that it can be attributed to the type of additives each manufacturer uses to add to improve fuel's lubricity as well as to the actual polyaromatic content which again varies from one manufacturer to another. However, there is no clear consensus in the literature in regards to this issue.

Distillation temperature of GTL is higher than that of diesel when distillation percentage reaches below 70% or lower than when the percentage is above this limit. The T90 (90% distillation temperature) of GTL is lower than that of diesel by 6-7% [90], which indicates that GTL has less heavy distillation and is easier to evaporate and to form into a more combustible air-fuel mixture.

Aromatics, whether total or polyaromatics of GTL are much lower than conventional diesel. They are about 96% and 94% respectively lower than diesel fuel. The very low content of aromatics is inherited in the low-temperature F-T processes. This makes GTL fuel superior to conventional diesel with respect to smoke and PM emissions as will be shown in the subsequent subsection.

Cetane number of GTL fuel, which is a key property in the assessment of ignition quality and ignition delay, is significantly higher than that of conventional diesel. Its rating goes as high as the 80s and it varies remarkably from one manufacturer to another. Because of the very high n-paraffin content, F-T diesel, hence GTL diesel, exhibits high cetane numbers. GTL brands whose n-paraffin content is lower would certainly exhibit a lower cetane number. In the engine community, it is customary to use cetane index instead of cetane number which is an empirical determination that uses the fuel's distillation characteristics among other properties to reflect the fuel's cetane quality. It is worth noting here that one should refrain from using the cetane index rating in the case of GTL since it is yet to be determined whether these calculations are applicable for unconventional fuels such as F-T diesel [88].

Lower Heating Value of GTL also tends to be slightly higher, about 1.5 % higher than conventional diesel. This would have influence on the brake fuel consumption as would be illustrated in the later subsection.

The Hydrogen-to-Carbon ratio (H/C ratio) of GTL is about 13-15% higher than conventional diesel (weight ratio). This could be attributed to the composition of GTL fuel consisting mainly of normal and iso-paraffins. Because of the higher H/C ratio GTL fuel tends to produce less soot.

As already mentioned, actual properties of GTL diesel fuel are manufacturer-dependent and are, as a result, production method-dependent. It is important, therefore, to refer to the source of GTL fuel and its actual properties whenever power and emission performances are concerned.
In this work, all investigations were conducted using GTL fuel of SHELL Bintula, referred to as Winter SHELL GTL whose key properties are listed in table 2-1 with reference to conventional diesel.

	Diesel	GTL
Density @ 15°C (kg/m^3)	839.2	776.2
Cetane number	~52	74.6
Sulfur (ppm)	~403	<<10
Total Aromatics (wt%)	~27.7	~1.4
Polyaromatics (wt%)	~6.2	~0.4
50% distillation (°C)	~265.3	263.3
90% distillation (°C)	~330.8	297.4
Viscosity @ 40°C	~2.665	2.453
C (wt%)	~86.32	85.14
H (wt%)	~13.32	14.81
Lower heating value (MJ/kg)	~42.9	43.69

Table 2-1: Shell's GTL fuel properties

2.4.2 Impact of GTL Fuel on Diesel Combustion

It is a well-established consensus that fuel properties affect combustion enormously. In particular, those properties that have direct effect on mixture formation and ignition characteristics. The first property of such is the cetane number which is not only a measure of the ignition quality and completeness but, more importantly, a measure to the so-called ignition delay. Ignition delay, as already explained, determines the time available for air-fuel mixing and its length enormously affect mixture formation. Although quantification of the effect of cetane number on ignition delay is yet to be fully developed and well documented, the few attempts to do so give a reasonable insight on the matter. Actually, the issue of effect of cetane number gained momentum as the interest in oxygenated fuels grew. Ming Zheng et.al. [93], who performed measurements of ignition delay of high cetane number fuels and oxygenated fuels (biodiesel), in particular, showed an Arrhenius type dependence of ignition delay on the cetane number. An expression for the ignition delay was given in which the activation energy term was modified in such to account for the cetane number. Such expression is given in equation 2-11.

$$\tau_{id}(ms) = A\left([O_2]+[F_o]_{fuel}\right)^{-k}(p/bar)^{-n} \times \exp\left(\frac{2100\,\mathrm{K} \times [71.3/(CN+255)]}{T}\right) \qquad (2\text{-}11)$$

where CN is the fuel's cetane number.

From this correlation, it can be deduced that due to its high cetane number, GTL fuel exhibits a shorter ignition delay than conventional diesel which results in a shorter residence time available for mixing. This in return decreases the portion of premixed combustion which on one hand could increase sooting tendency but on the other decrease NOx emissions as well as decrease the steep rise in temperature and pressure and produce lesser combustion noise.

The lower distillation temperature of GTL fuel, on the other hand, improves its evaporation characteristics and, therefore, somewhat balances the cetane number effect on mixing formation. Because of its lower distillation temperature, GTL fuel exhibits on faster evaporation rates, on the one hand, while on the other develops a shorter liquid jet length [94].

Furthermore, due to the lower density compared to conventional diesel fuel, GTL fuel exhibits fuel jets with lower fluid inertia and, thus, fuel jets with shorter liquid length [94]. Because of the lower density, the bulk modulus of compressibility is also lower. This makes GTL fuel more compressible than conventional diesel so the pressure in the fuel injection system develops is slower and the pressure waves propagate slower which, depending on fuel injection system type, could lead to a later injection timing (or longer injection delay) than conventional diesel [95,96]. In this case, the locations of peak pressure and peak Heat Release Rate [HRR] are slightly later than those of diesel as observed by Tao Wu et. al. [97].

The higher Lower Heating Value [LHV] of GTL fuel yields better fuel economy than diesel. The Brake-Specific Fuel Consumption [BSFC], as a result, is about 3% lower than diesel [97]. This contributes enormously to the favoring of GTL fuel over conventional diesel. It is worth noting, however, that the Brake Thermal Efficiency [BTE], although higher than diesel at all engine conditions [97], is not proportional to BSFC. This can be explained if considering the BTE formula shown in equation 2-12

$$\eta_{bt}[\%] = \frac{3.6\times 10^{6}}{BSFC \times LHV} \times 100 \qquad (2\text{-}12)$$

According to this equation, BTE is inversely proportional to the product of BSFC and LHV. So for the case of GTL, who has lower BSFC and higher LHV, the end product levels off somewhat.

Given the growing interest in smokeless combustion and ultra-low NOx emissions, new concepts are emerging other than Homogenous Charge Compression Ignition [HCCI] which exhibits high HC and CO emissions. The Modulated Kinetics [MK] combustion or Low Temperature Combustion [LTC] (they are related to large extent) is becoming more attractive to engine applications. However because of the ultra-high EGR rates required to lower cylinder temperature combined with low cetane number fuel which together prolong ignition delay to times long enough to achieve premixed combustion, thermal efficiency decreases and HC emission increases [98]. As a result GTL is emerging as good replacement of the conventional diesel because of its superior ignition quality.
Kawamoto et. al. [99] have found that increasing EGR rates is insufficient to achieve MK combustion since it does not guarantees reducing NOx emission to the desired level. As a solution Kawamoto et. al. have proposed the combination of low-compression ratio and GTL fuel for the realization of MK combustion over a wider engine operation range. Such an approach guarantees an ignition delay long enough to achieve fuel injection completion before onset of combustion and due to its high CN and high paraffin compounds, realization of LTC and premixed combustion was possible while inhibiting HC increase even at cold conditions.

2.4.3 Impact of GTL Fuel on Pollutant Emissions

Pollutant emissions of diesel combustion is a response to the combustion system, encompassing injection system and strategy, and fuel type employed. The impact of fuel type is brought about by the effect of fuel properties on exhaust gas emissions. Here, the effect of key fuel properties distinguishing the GTL fuel on pollutant emissions shall be demonstrated.

Effect of Aromatics seems to be the most disputable among other properties. Some research work has shown an increase in soot formation with an increasing amount of aromatics [100,101,102, 103] and, thus, GTL fuel because of its low aromatic content bears reduction in soot formation. Other research work, however, showed that aromatics have a minor effect on the soot formation [104] and the reduction in soot and PM in general brought about by employing GTL fuel has been attributed to a higher cetane number [105] and higher H/C ratio [106].
The reason for such disagreement in the various research works seems to be the decoupling of the effects of various properties. Those who reported aromatics to have an effect on soot reduction did not hold cetane number and ignition delay constant.

Investigation of the impact of GTL fuel conducted on commercial diesel engines did show, however, that the low-aromatic and low-PAH content of the GTL not only reduced PM but also HC and NOx emissions [107]

Because sulfur in the fuel is oxidized to SO_2, which can combine with unburned hydrocarbon and become absorbed by neighboring soot particle, the near-zero sulfur content in GTL fuel contributes to a large reduction in PM emissions [104,107]. As a matter of fact, because of its low-sulfur content the employment of GTL can enable advance after treatment technologies. For example, the near-zero sulfur content lessens the corrosion of the EGR component brought about by the sulfuric acid. Furthermore, the performance of the diesel oxidation catalyst is also improved since the oxidation of SO_2 is inhibited resulting in less PM produced in the catalyst. Interestingly, the near-zero sulfur content of GTL fuel also improved the durability of the NOx adsorber catalyst. The effect of fuel sulfur on emission control devices has been well-documented and the reader is referred to reference [108] for further details.

Cetane number has a double effect on pollutant emissions. It shortens ignition delay as well as the lift-off length of a jet-flame resulting in an increase in the tendency to form soot [109]. Soot formation levels of GTL are, however, much lower than conventional diesel fuel. Low aromatic and low-sulfur effects seem to overtake the cetane number effect [94,89]. On the other hand, because of its effect on ignition delay and the portion of premixed-combustion due to a higher cetane number, GTL fuels reduced NOx emission remarkably [109,107].

The higher hydrogen to carbon ratio of the GTL fuel contributes significantly to the reduction of soot formation. Although the relative importance of H/C ratio in comparison to other properties is yet well-established, the lower carbon content contributes to reduction in soot and PM emission in general.

The lower distillation temperatures of the GTL fuel, mainly T90, seem to also promote further reduction in soot and PM emissions. This has been observed in several works of research [89.91,94] which related faster evaporation tendencies to air-fuel premixing enhancement.

The various GTL fuel properties all contribute to produce simultaneous reduction in PM, NOx, HC and CO emissions [88,89,90,94]. This has been largely demonstrated from investigation conducted on light and heavy duty commercial diesel engines.
The Ad Hoc diesel Fuel test program [110], consisting of tests performed on Ford, GM and Damilerchrysler engines, showed that without EGR, GTL fuel reduced PM and NOx emissions compared to Low Sulfur highly Hydrocracked [LSHC] fuel.

Johnson et. al. [111] performed investigations using Peugeot 405 light-duty diesel engine with indirect injection and revealed the superiority of F-T fuels (GTL mainly) over other low-sulfur diesel fuels. They have reported, however, that the T95 temperature had little influence on the PM emissions from F-T fuels.

Tests conducted by Sasol Oil on heavy duty DDC series 60 diesel engine [112] showed that neat GTL fuel had the largest emissions reduction for all regulated emissions.

Further studies were performed to examine the emission performance of heavy-duty trucks [113] and transit buses [114] operating on GTL fuels. Depending on engine age GTL fuel reduced the NOx emission by 5-12%, PM by 24-31%, CO by 18-49%, and HC by 20-40%.

The recent investigations conducted by Tao Wu et. al. [97] in commercial medium size six-cylinder DI diesel engine operated on GTL fuel showed that GTL reduces CO, PM, HC and NOx emissions at all loads while the greatest advantage of GTL has been observed at high loads. GTL reduced CO emissions by 26.7%, HC by 20.2%, soot by 15.6% and NOx by 12.1%.

The joint SHELL-BOSCH-VW research work [115] on the impact of GTL fuel on engines configured to the EURO 4/5 emission standards showed that neat GTL fuel (referred to as SMDs) demonstrated great benefits in PM and NOx emission reduction as well as remarkable improvements in fuel consumption compared to other sulfur-free diesel. A summary of the results are shown in figure 2-17

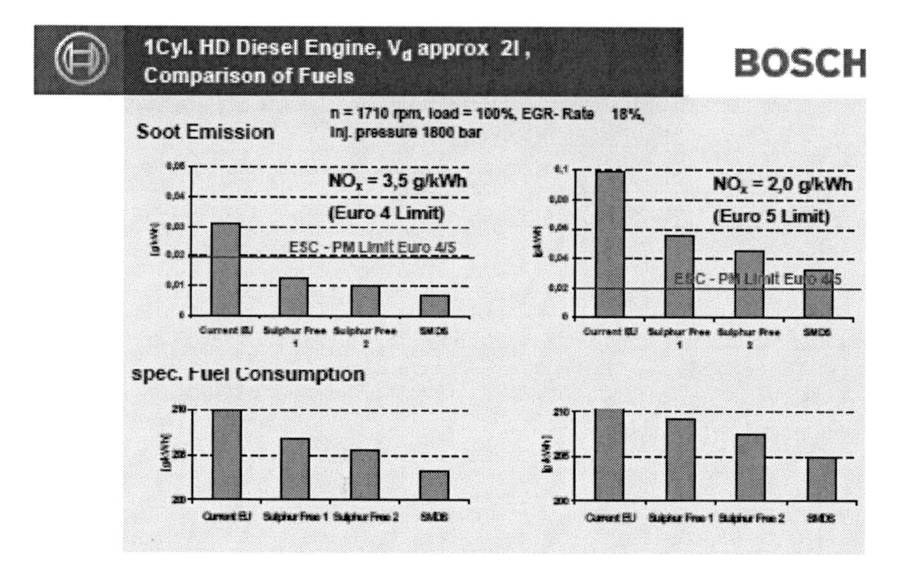

Figure 2-17: Results of SHELL-Bosch-VW joint research on GTL emission and fuel consumption performance in comparison to other sulfur-free diesel fuels. From Ralph [115]

In summary, the examination of the impact of GTL fuel on pollutant emissions and engine overall performance, as already described, were conducted on existing engines with minimal modifications.

The majority of these investigations were comparative in nature, comparing GTL performance to conventional diesel and other alternative fuels. Attempts to configure diesel engines to fully exploit advantages of GTL fuel stemming from their unique properties have not yet been fully explored. Driven by the conclusions of the works by Kawamoto et. al. [99] and Kitano et. al. [89], this work places emphasis on the employment of low compression ratio, which in combination with neat GTL fuel should yield the prerequisites for advance combustion concept based on LTC and premixed combustion. The core of this work is, therefore, to examine the effect of injection parameters and injection strategies on the soot formation in a low-compression ratio DI diesel engines operated on neat GTL fuel.

3 Experimental Set-Up

3.1 Rapid Compression Machine

All experimental investigations conducted in this work were performed using the Rapid Compression Machine [RCM], an optically-accessible single cylinder machine that is capable of simulating a compression stroke and partially an expansion stroke of a real engine. The RCM offers several advantageous. A flexible adjustment of operating conditions and flexible adjustment of compression ratio as well as multiple optical accesses of the combustion chamber and easy-to-modify piston and cylinder head.
A layout of the machine is shown in figure 3-1 while a detailed description of its parts is provided in appendix I.

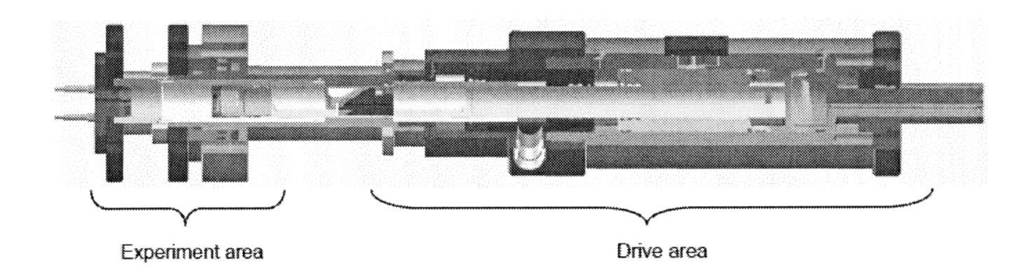

Figure 3-1: Layout of the RCM. From TESTEM, RCM manufacturer's manual

The experimental area consists basically of all parts associated with the compression and combustion chamber. A uniquely designed piston with a silica glass inserted in it provides an axial optical access allowing visualization of the combustion chamber from below projecting the plane perpendicular to the cylinder reciprocating direction. Although the glass insert can be designed to any desired geometry, the one that has been used in this work is a flat glass generating as of a result flat piston bowl (mould). Its glass properties are of those capable of transmitting a wide spectrum of light wavelength including the ultra-violet range. The cylinder-wall on the other hand consists of two parts, a cylinder cartage and ring-shaped glass insert allowing radial inspection of the combustion chamber. This optical access allows visualization of plane parallel to the cylinder reciprocating direction as well as side illumination.
The cylinder head is the part in which the diesel injector and pressure transducer are installed. The pressure transducer which provides time-resolved cylinder pressure measurements is of the Kisler 6061B type. The inner surface of the head is fitted with special damping ring to avoid crash or slamming of the piston onto the head. Both cylinder-wall and working piston possess as heat sink given their bulk metal mass.
To better simulate the thermodynamic state of a real engine both the working piston and cylinder wall are heated through an electrical heaters installed in the piston crown housing and cylinder wall respectively.

The temperature to which both piston and cylinder-wall are heated is controllable and has been set in this work to 60°C (higher temperature was avoided to avoid undesired heating instabilities).
The driving area of the RCM houses the mechanism which generates the rapid movement of the piston during compression. Hydraulic oil is used given its incompressibility as the working fluid which exerts the desired loading on the so called piston road to generate the desired acceleration. The working principle of the rapid compression stroke is illustrated in figure 3-2.

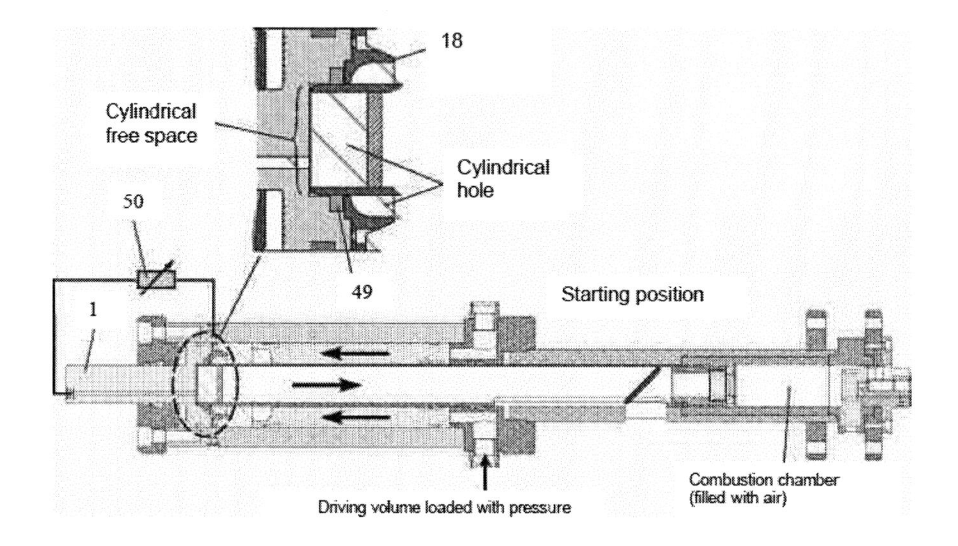

Figure 3-2: Illustration of the working principle of the compresion stroke of the RCM. From TESTEM RCM manufacturer's manual

Two coaxially-mounted cylindrical balance masses piston (marked 17 in the layout in Appendix I) move simultaneously yet oppositely to the movement of the working piston-rod assembly. Compressed air which provides the driving force is fed upstream these two pistons while hydraulic oil that is located in the coupling volume which connects the two volumes of the balance and working piston transfers momentum to the working piston and induces piston motion. When the working piston is at Bottom Dead Center [BDC], hence at the zero point (starting point) of the stroke, the hydraulic oil that under pressure loading exerted by the compressed air upstream can not flow into the driving volume of the working piston thanks to the radial sealing (marked 49 in the enlarged schematic drawing of Figure 3-2) positioned directly at the coupling volume.
Thus, no movement of both working and balance piston is possible. Upon activating the bypass valve (marked 50 in figure 3-2) hydraulic oil flows from the cylinder holes through the bypass tube and starts pushing the working piston upward (left to right).
The working piston moves slowly until the cylinder holes of the coupling volume are uncovered allowing larger flow mass of the hydraulic oil to flow in. As a result the working piston accelerates and moves upward rapidly.

The opposite motion of the balance pistons guarantees that the acceleration of the working piston translates in vibration-less motion. A critical feature needed for smooth operation of optical diagnostics.
Towards reaching Top Dead Center [TDC] the cylinder pressure increases and constraints the working piston motion. This damping effect yields deceleration of working piston. To make this deceleration a controllable process, hydraulic oil is flown into control volume upstream the coupling rod upon activation of a throttle valve and slows the motion of the working piston. This effect is shown schematically in figure 3-3.

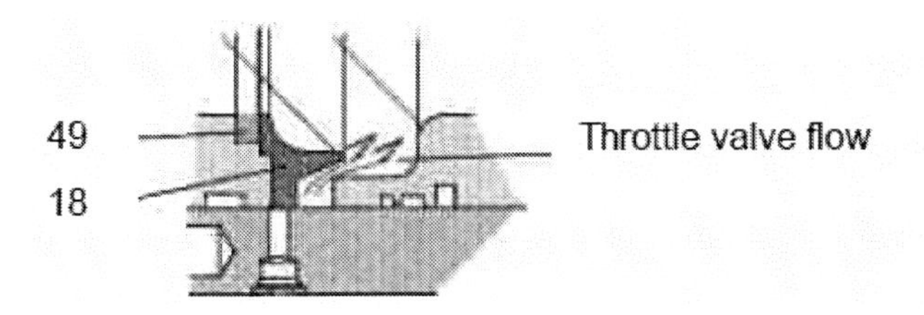

Figure 3-3: Illustration of piston deceleration. From TESTEM RCM manufacturer's manual

For optimal measurements and assessment of the temporal thermodynamic state of the combustion chamber, proper determination of compression stroke and TDC is needed. The controllable piston deceleration mechanism the RCM offers allows proper determination of the two. During calibration cycle the working piston is accelerated towards the head at low driving force while regularly activating the throttle valve until further incremental movement upward results in negligible cylinder pressure increase. Thus the TDC point at the desired stroke length is determined.
Furthermore, the rate at which hydraulic oil flows into the coupling volume to drive the working piston determines its acceleration. An adjustable inlet width allows the control of the cross sectional area through which oil flows and thus the flow rate is altered. Such adjustment allows setting piston acceleration with a desired correlated engine speed. In this work piston acceleration has been set to correlate to an engine speed of 1500 RPM. This has been kept unchanged throughout the entire experiments.

An incremental lineal piston positioning sensor shown in figure 3-4 is positioned within the so called connecting rod and gives an instantaneous positioning signal of the working piston.
This allows the determination of the actual stroke and its time-resolved signal is used to first provide the time-resolved cylinder pressure and volume measurements, and second to provide piston position-based triggering tool. Thus, triggering of key events such as fuel injection and optical visualization or illumination devices is done on piston position-basis.

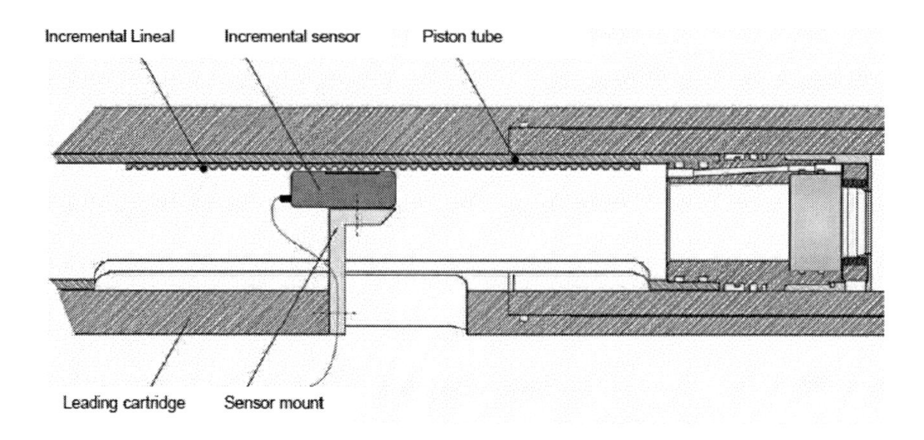

Figure 3-4: Illustration of the lineal piston-positioning sensor of the RCM. From TESTEM RCM manufacturer's manual

The RCM control system also offers data acquisition tools. It provides 4 channels signal acquisition and processing. The time-resolved signal of the Kistler pressure transducer is first amplified through the Kistler charge amplifier, type Charge Moter 5015, and then fed to the RCM Data Acquisition System [DAS], similarly is the signal from the piston position signal. The remaining two channels are used for feedback trigger signal from injection and camera control unit. Thus the RCM DAS provides text file of cylinder pressure and volume which are later used for the cycle thermodynamical analysis.

All operating procedures of the RCM are computer-controlled. For further information and description the reader is referred to the operating manual of the RCM provided by the manufacturer TESTEM available at www.testem.de as well as to the design work of Ofner [46].

3.2 Injection System

The injection system deployed in this work consists of 1- commercial Siemens VDO (now became Continental Automotives) Piezoelectric injector of the series PCR2 installed typically in Renault diesel engines, 2- commercial High Pressure Common Rail Pump of the series A2C20000754AAC K9K EU4-Engine, and 3- Research injection control unit.

Piezoelectric injector of the type PCR2 made by Continental Automotives has been chosen as injector of choice given its unique characteristic of fast needle opening and closing. This is particularly advantageous for multiple injection specially in the cases of close injection events as deployed in this work. The piezoelectric injectors are superior to solenoid in many other aspects as those described in section 2.3.2.

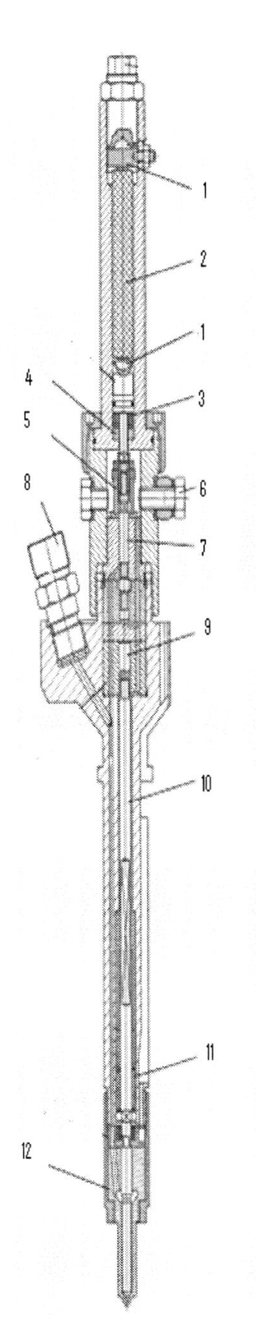

A layout of the piezoelectric injector is schematically shown in figure 3-5. It was operated on low-frequency pulsed activating voltage (charge). Before activation the injector is in normally-closed position during which the injector actuator is continuously loaded thus holding the Piezo-element (Piezo-christal) in stretched (elongated) position. Upon needle opening, the actuator is momentarily discharged resulting in contraction of the Piezo-element. Recharging the actuator expands the piezo-element and closes the injector needle. Since needle activation and deactivation occurs in an finite time, a minimum dwell time between end of one injection event and a start of subsequent one is 100μs (longer time interval is recommended though to avoid poor hydraulic flow). Such relatively short time allows very close subsequent injection events. In this work, however, two subsequent injection events were separated by times longer than 300 μs to secure stable hydraulic flow given that fuel was pressurized for short periods because of the cyclic operation of the RCM. An ultra high-speed shadowgraphy imaging at a rate of 50,000 fps using the HS-camera has been deployed to determine the time period between $SOI_{electric}$ and $SOI_{hydraulic}$ in the pressure range up to 1600 bar. This has been found to be around 320 (±10)μs and almost independent of pressure.

Injection timing was controlled through the piston-position RCM-generated trigger signals while injection duration was commanded through an external signal generator, type Stanford Research Signal generator DG535. In the case of multiple injection, a multi channel multiplexer was used who receives multi injection trigger signals as input and transmits a single command signal containing all injection events to the single channel trigger input of the injector control unit. The injector nozzle is of 6 holes whose orifice angle and diameter are 150° respectively.

Figure 3-5: Layout of typical Piezo-electric Injector; 1- Bearings, 2- Piezo-actuator, 3- Spring pack, 4- Balance ring, 5- Valve sprig, 6- Nipple for fuel return flow, 7- Valve needle, 8- High pressure fuel supply connector, 9- Actuator piston, 10- Plunger rod, 11- Nozzle spring, 12- nozzle

High Pressure Pump [HPP] deployed in this work is of the type of three cylinder-radial piston pump with a displacement of 0.8 cm^3 / rpm and maximum fuel pressure of 1600 bar. A layout of the HPP is shown in figure 3-6.

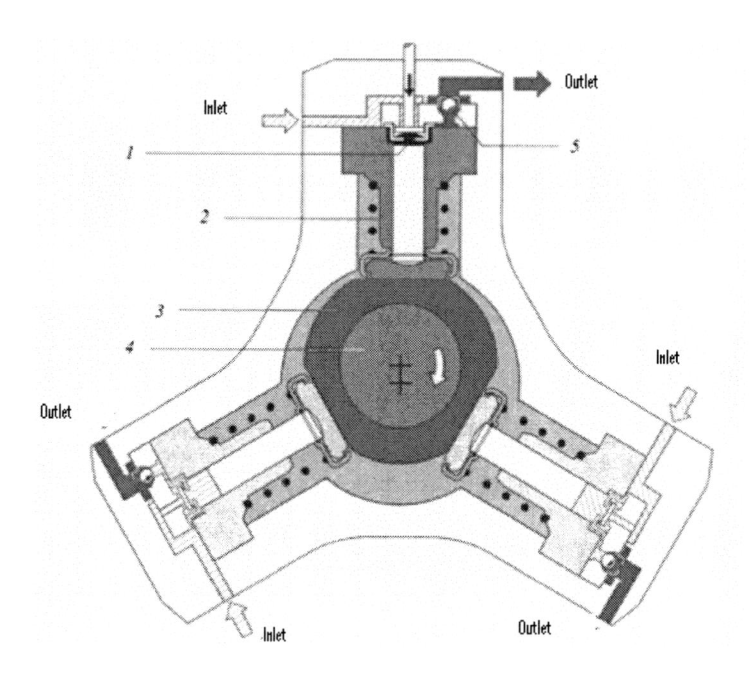

Figure 3-6: Layout of the HP Pump. 1- Intake valve, 2- Pumping element with pumping piston, 3- Eccentric cam, 4- Working shaft, 5- outlet valve.

According to manufacturer instructions, the maximum water content in fuel allowed is (200 ppm). Something that did not require any action since neither GTL fuel nor its storage and pumping at the laboratory conditions could produce water content close to this limit. The pump was operated through an external laboratory electrical motor and the pressurized fuel is transferred to four-cylinder serial common rail and then to the injector installed in the RCM head through a high-pressure supply line. Pressure variation was controlled electronically through control current of the Pressure Control Valve [PCV] of the pump governed by an external commercial laboratory control unit.

Injection control unit used in this work was a research model built and provided by Continental Automotives called RESI which was specifically tuned for the operation with the PCR2 injector. It receives a trigger as its input signal while transmits a single needle activation signal as its output signal.

4 Measurement Techniques

4.1 Spray Visualization Techniques

Fuel introduction into combustion chamber and its associated mixture formation dynamics are of paramount importance as pertains to combustion characteristics and pollutants formation. Obtaining information on in-cylinder spray development, its evaporation pattern and the subsequent ignition zones and ignition characteristics is, therefore, very critical to the understanding of combustion and pollutant formation of a given combustion system and fuel injection strategy. Optical diagnostics have become, as a result, very attractive tools for the realization of such objective and several techniques have been developed ranging from straight forward and simple-to-apply to very complex depending on the type and accuracy of measurements. Most of the techniques applied to real combustion systems such those applied to diesel engine environment are, although of various complexities, qualitative to semi-quantitative in nature. This is because of the harsh environment of a diesel combustion system as well as because the complex structure of diesel fuel that is often regarded as multi-component liquid. Certainly, the more complex the technique is, the more sophisticated yet expensive the equipment and devices deployed are. From several decades of experimentation it has been found that simpler techniques can provide more than 80 % of the knowledge that other more complex techniques can provide. It is acceptable, therefore, not to hesitate to apply simpler techniques, particularly when investigations of the comparative nature are involved.

4.1.1 Combined LIF-Mie Technique

The combined LIF-Mie technique is applied for the simultaneous detection of liquid and vapor phase of a diesel fuel which can provide qualitative assessment of the spray development and temporal as well as spatial mapping of fuel evaporation and ultimately fuel-air mixing. The Laser-Induced Fluorescence [LIF] is deployed for the detection of the liquid as well as vapor phase of the fuel while Mie is deployed for the detection of the liquid phase only. When comparing the simultaneous acquisition of the two, one can decouple the evaporated from the non-evaporated zones and follow mixture formation pattern and its association to combustion.

The LIF measurement is basically based on a fluorescence emitted from a molecule that is at excited-stage which has changed electronic level. Depending on the excitation source, fluorescence can be narrow or broad-band. The later yields a broad fluorescence spectrum and is suitable for the detection of hydrocarbon bonds of a real fuel through which one obtains similar emission spectrum for both liquid and vapor phase (for detailed description the reader is referred to the literature Eckbreth [122] and Mayinger and Oliver [123].)

The Mie measurement, on the other hand, is based on elastic light scattered from an object. Theoritically, object is treated as perfectly spherical. Since this is typically not the case in real systems quantitative measurements as could be obtained from the Lorenz-Mie theory is impossible. Mie measurements are, therefore, qualitative in nature.

In practical applications the wave length of the light source deployed is significantly bigger than the object (particle) size. This yields light scattering in the Mie regime and normally referred to liquid phase.

Since light is scattered from non-fully spherical object (particle) scattering is not even in all directions. Matter of fact, intensity distribution of scattered light shows that intensity varies with scattering direction. It seems that maximum light intensity occurs at scattering angle of 120°. In an experimental system like the one deployed in this work, hence the RCM, the optical access is at 90° with the scattering plane. Although intensity at this angle is not the highest, the Mie signal intensity is still strong enough and can be easily detected. Since Mie is an elastic process no energy loss is encountered. The Mie scattered light is, therefore, of the same wave length as the incoming light source.

For the simultaneous LIF-Mie measurement a light source that should be deployed is a one that is of sufficient energy to first excite diesel fuel molecule as well as to yield a scattered light of sufficient intensity. An Excimer UV laser operated with XeCl for the generation of excitation wave length of 308 nm has been used. This wave length is a favorable one because it can excite the abundant aromatic bonds with one or two rings in diesel fuel. The fluorescence which corresponds to both liquid and vapor phase, is as of a result, broadband and usually in the 350-450 nm range. The Mie signal is, on the other hand, of the same wave length as the laser, hence 308 nm. In order to decouple or separate the Mie signal from the LIF signal, a dichroic mirror has been mounted which reflects lights of wave length smaller than 350 nm and transmits those of wave length higher than 350 if positioned at 45° as shown in the schematic layout of the entire optical set-up in figure 4-1.

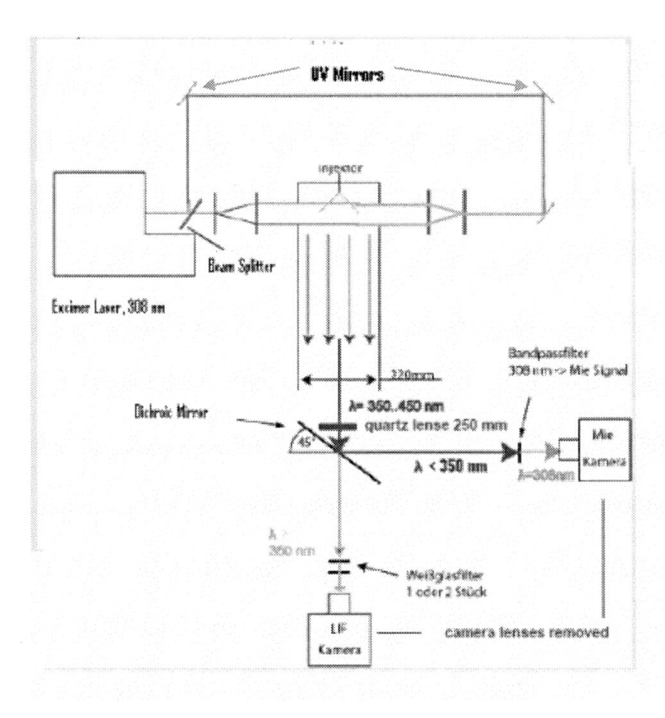

Figure 4-1: Layout of the combined LIF-Mie optical set-up

The reflected light which contains the Mie signal passes further through a bandpass filter centered at wave length of 308 nm and then is acquired through an intensified CCD camera. The transmitted light, on the other hand, which corresponds to the fluorescence signal, passes through a broad band filter of the range 350-450 nm and is acquired by a second intensified CCD camera. In order to capture larger portion of a spray the laser light has been expanded and formed into a laser light beam instead of a sheet. Although this leads to an attenuation of the laser, this was necessary in order to reduce the number of measurements and cycles required to map the entire spray given the low operating frequency of the Excimer laser which is of 10 Hz. This method has also been adopted by Henle et. al. [116] and Uhl et.al. [117] and because of the low operating frequency of the laser only one Mie and LIF image could be acquired in every cycle. Furthermore, given that diesel fuel particle absorbs the incoming light, the laser beam has been split through a beam splitter into two beams, each was directed through set of mirror to illuminate the combustion chamber from both sides.
An example of LIF-Mie images acquired for a conventional diesel fuel and using an RCM other than the one described in chapter three which has an axial optical access (as depicted in figure 4-1) is shown in figure 4-2.

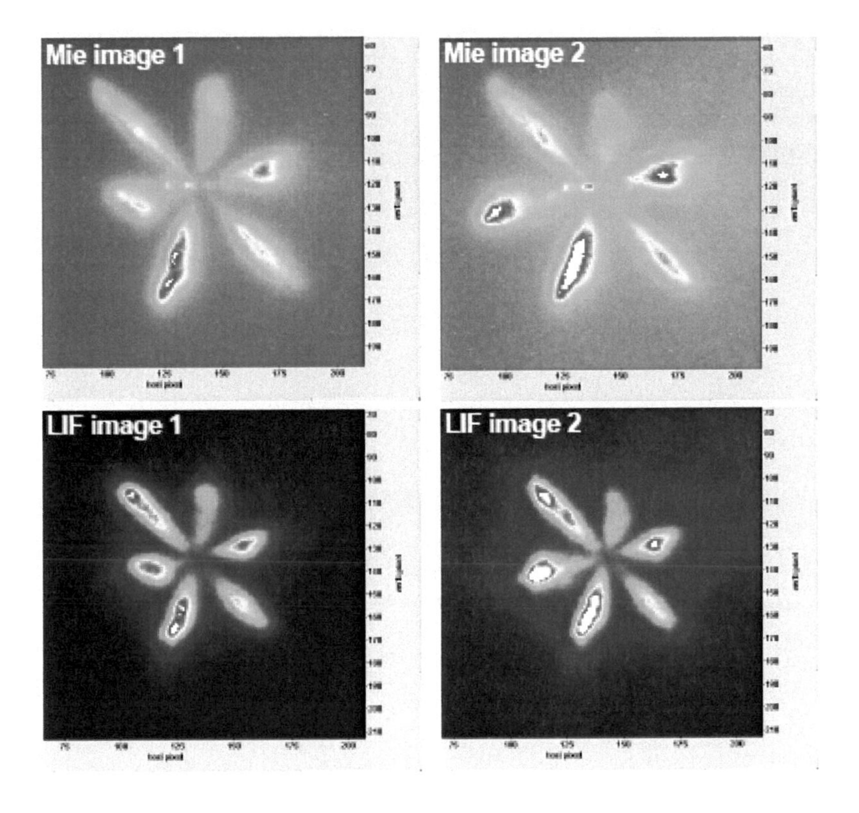

Figure 4-2: Exemplary LIF-Mie images of diesel spray

Figure 4-2 shows inhomogeneous illumination of the combustion chamber something that could falsify measurements. This could be attributed to the Excimer laser whose intensity exhibits non-even temporal and spatial distribution mainly because of the elongated optical paths as well as the method of lasing. This alongside the low operating frequency of the laser impeded the conduction of reliable and reproducible measurements at the RCM particularly because of the cyclic operation of the RCM by which thermodynamic states could not be maintained. Most importantly, applying this technique for the case of GTL fuel was not fully feasible since there has been no proof that the excitation wave length of 308 could also be used to excite GTL fuel molecule given its poor content of aromatic bonds which impeded the interpretation of the fluorescence signal. Determination of the proper or optimal excitation wave length of GTL fuel was beyond the scope of this work.
Furthermore, since temporal and spatial assessment of fuel evaporation and mixture formation was by no means the core of this work, other technique that supply satisfactory info on spray development, yet much simpler to apply, was adopted instead. This would be described in the following sub-sections.

4.1.2 Schlieren / Shadowgraph Techniques

The Schlieren technique is used to visualize medium of inhomogeneous refraction index or in physical terms of inhomogeneous density through which the detection of liquid and vapor phases is made possible. If light passes through a transparent test medium then the light rays would bend in proportion to the local refraction index and would exist the test volume with a displacement (deflection) proportional to the local refraction index gradients expressed in the form of

$$\varepsilon_x = \frac{L}{n_0}\frac{\partial n}{\partial x}, \varepsilon_y = \frac{L}{n_0}\frac{\partial n}{\partial y} \tag{4-1}$$

where L is the schlieren extent and n_0 is the refraction index of the surrounding medium

As a result regions of different refraction index would project the bent light rays at different distances allowing thus the detection and separation of regions of different refraction index. The schlieren imaging is based, therefore, on such separation principle which by adding a blocking element (usually referred to as schlieren knife for one dimensional separation and schlieren aperture for two-dimensional) the bent light rays are selectively blocked resulting in dark spot on the imaging plane while the unblocked rays are allowed through creating a bright spot. Thus imaging of the desired test object is translated in terms of dark points in the background of bright points. The selection of which bent light rays to block usually referred to as *cut-off ratio* which defines, also, the overall system sensitivity. System sensitivity conflicts, however, with the intensity of the light source since by increasing cut-off ratio lesser light is allowed through to provide bright background or in other words, image contrast. Higher light intensity is required to compensate such loss in image brightness and sensitivity.
Schlieren can be applied through either lens or mirror system. Mirror system is, however, more common and usually of better performance [118]. A typical schlieren system for transparent test medium is the so called Z-Type shown in figure 4-3.

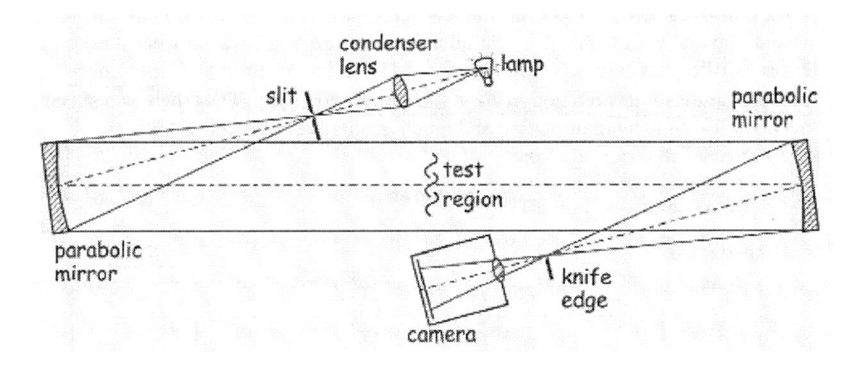

Fig 4.3: Z-type schlieren arrangement. From settles [118]

Given the separation principle on which the schlieren method is based, hence bent light rays are displaced in proportion to the refraction index with respect to the horizontal optical axis, the incoming or illuminating light is desired to be parallel which is realized through the mounting of a parabolic mirror while a point light source is positioned at its focal point. (in case a readily available point light source of a sufficient intensity is not available, point source can be realized by a system of non-point light source, condenser (focusing) lens and a slit or another schlieren aperture). A second parabolic mirror is used to refocus the projected bent light rays to a point located at this mirror's focal point but on the opposite side of the optical axis, as shown in figure 4-4, where the schlieren aperture (or knife) is positioned. The two parabolic mirrors are positioned on opposite side of the axis in order to cancel any undesired coma [118]. Upon determining the proper cut-off ratio (the opening of the schlieren aperture) the imaged object is projected onto a camera which in return acquires the image.

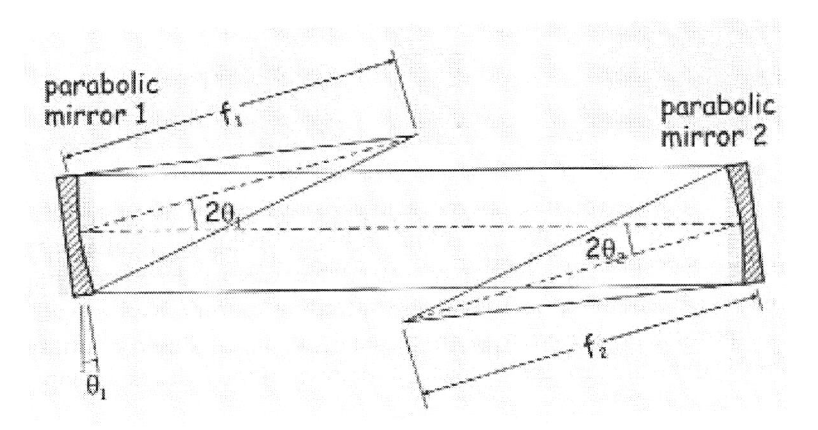

Figure 4-4: Diagram of Z-type schlieren system emphasizing positioning of parabolic mirrors. . From settles [118]

Practically speaking both parabolic mirrors are desired to have similar focal point so the set-up is fully symmetric (this has also economical advantage) and they both are tilted with angle θ_1 and the distance between the two mirrors are desired to be of 2f [118]. To avoid or minimize aberration the tilt angle is desired to be small while the mirror's focal point is desired to be long (the later improves system sensitivity as well).
It is quite accustomed to define schlieren system sensitivity in terms of minimum detectable gradient which is expressed in terms of cut-off ratio a, field distance L and the focal point of the parabolic mirror as depicted in equation 4-2 [118].

$$\frac{\partial n}{\partial x}_{\min} = C\frac{n_0 a}{Lf} \qquad (4-2)$$

where C is a contrast threshold

Thus the longer the focal point of the parabolic mirror and the smaller the cut-off ratio are the higher the system sensitivity. A key factor which plays a major rule however, is intensity of light source as already mentioned.

In practical systems, however, like the one used in this work, hence the RCM, the test object (fuel jets) is not positioned in fully transparent test chamber and thus the projected bent light rays can not exist the test chamber from an opposite side. Rather, from the same side as that of the illuminating light. A planar reflecting surface is, therefore, used instead of a second parabolic mirror which reflects the bent light ray and forces it to exist from the incoming optical access. This means that the reflected light would pass a second time through the schliere something that usually generates double image. In the case of the RCM, the cylinder head has been carefully polished and served as a fully reflective surface. Given that the fuel jet existing the injector nozzle is positioned at a point almost at zero distance from the reflecting head, the effect of double image is almost cancelled.

The schlieren set-up used at the RCM is shown in figure 4-5. The single parabolic mirror has a focal point of 150 mm. Initially a CW Argon-Ion laser has been sought to be used as light source of high intensity but given its coherent nature, undesired speckles emerged as a result of rough surface roughness of the parabolic mirror. Instead a pulsed flash lamp which can generate flash period of up to 1ms has been deployed. An optic fiber light guide has been attached to the flash lamp in order to first flexibly transport light to the optical set-up and second to translate the light source to a point light source. A light guide of diameter of 0.5 mm has been first used during system alignment for optimal positioning of the light source at the focal point of the parabolic mirror for the generation of parallel illuminating beam. Later on, however, and during experimentation it has been replaced with light guide of the diameter of 1.5 mm to gain higher light intensity. Higher light intensity becomes crucial if the exposure time of the acquisition camera becomes shorter. Shorter exposure time is definitely desired in order to minimize the effect of motion blur of a propagating fuel jet.

Figure 4-5:
Schlieren Optical set-up

1- Flash lamp coupled to light guide

2- Parabolic mirror

3- RCM's reflecting mirror

4- RCM

5- Reflecting polished cylinder head

6- Analyzing beam reflecting mirror

7- Schlieren Cut-off circular aperture

8- Focusing lens

9- Acquisition HS CCD Camera

A beam splitter has been initially mounted in front of the light source so to separate the illuminating and the analyzing (projected) light. This in return wastes about three quarter of the captured luminance of the light source and thus reduces the effective light intensity dramatically resulting in a severe reduction of image contrast and system sensitivity. To avoid such undesired situation, the parabolic mirror has been intentionally yet slightly misaligned with respect to the RCM's optical axis so the light beam reflected from the RCM and contains the imaged schliere is slightly displaced from the illuminating beam. A very small circular mirror has been mounted to reflect the reflected beam away from the illuminating beam and directs it to the schlieren aperture plane. The schlieren aperture has been positioned at a distance equal to that of the circular mirror from the light point source (light guide holder). Thus it's guaranteed that the schlieren aperture is positioned at the focal length of the analyzing beam. Upstream the schlieren aperture a focusing lens has been mounted to focus the imaged schliere onto the High-Speed CCD camera which acquired images at a rate of 15000 fps and resolution of 256x256 pixels. The camera's objective has been removed as a result and its shutter has been set to provide exposure time of 10μs. An exemplary schlieren images of propagating fuel jets are shown in figure 4-6.

Several constraints have impeded, however, the full utilization of schlieren imaging for the analysis of spray development. First, given the relatively smaller test chamber of the RCM (about 80 mm in diameter), the injected spray fills rapidly the test chamber and the resulted local density gradient becomes rapidly very small which requires higher sensitivity for the schlieren system to be able to detect such gradients.

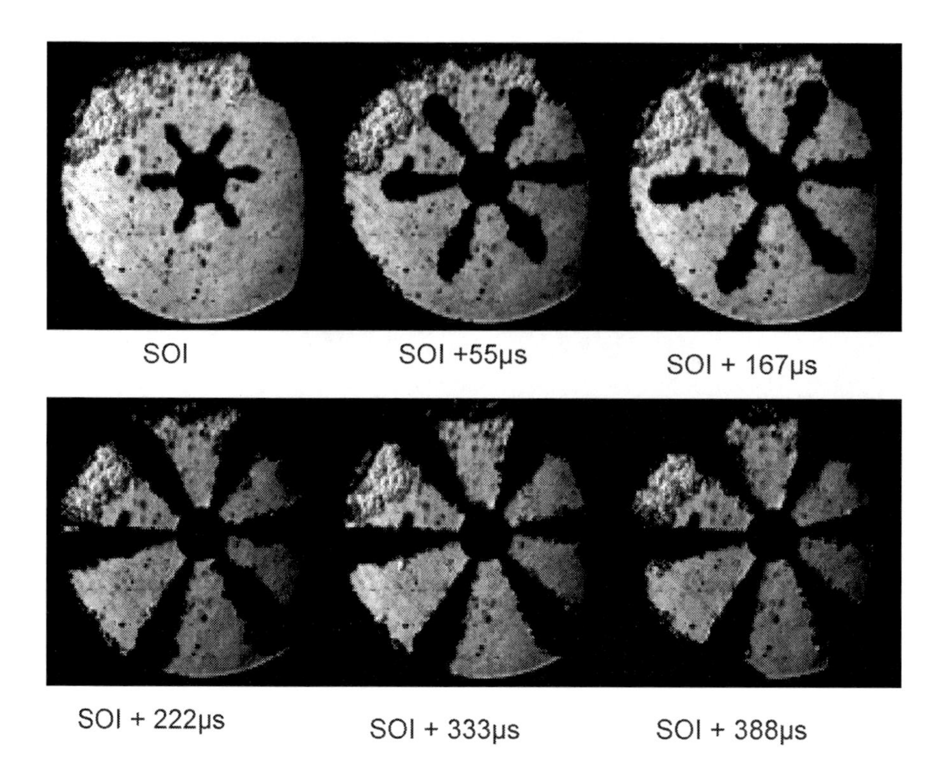

Figure 4-6: Sequence of Schlieren images of propagating fuel spray.

Further reduction of the cut-off ratio decreases light intensity dramatically and reduces image contrast to a non-visible level. This has been experienced also by Eisen [69] but was not an issue when a test chamber such that of a larger cylinder RCM (cylinder diameter of 220 mm) was deployed. Second, the parallel illuminating light has reduced the effective view field and as a result, only smaller portion of the propagating spray could be visualized. Third, visualization had to be conducted under compressed combustion conditions (air was replaced with nitrogen) because the reflection quality of the polished cylinder head deteriorated rapidly as a result of combustion. Multiple polishing has imposed tedious working conditions because it reduced the effective cylinder head thickness thus reducing the effective compression ratio. Ignition zones could not be simultaneously visualized. As a result, diverging shodowgraphy imaging has been deployed instead whose general optical set-up is shown in figure 4-7.

It is obvious how simpler the shadowgraph imaging set-up is with respect to the schlieren. A diverging light source, in this case, a halogen lamp is used as illuminating light source whose optical axis is intentionally misaligned with the light beam reflected from the RCM cylinder head to avoid using a beam splitter.

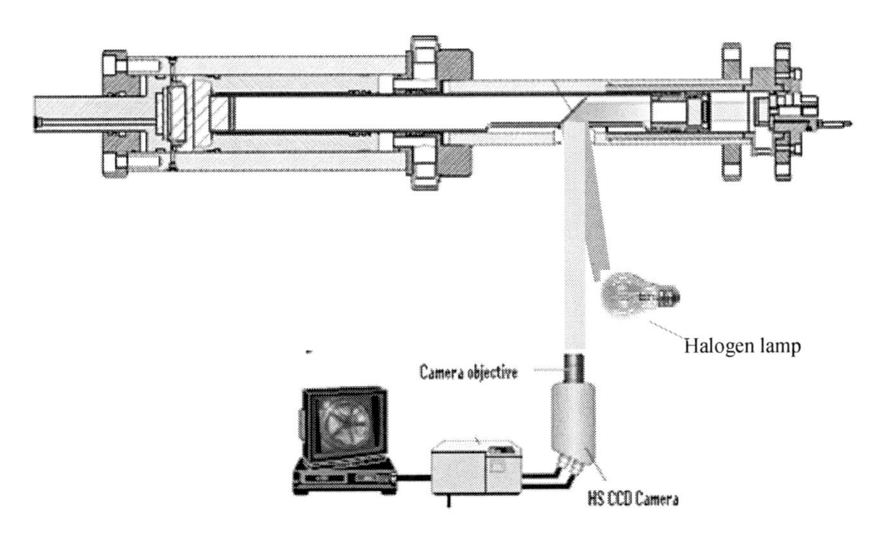

Figure 4-7: Shadowgraphy optical set-up

The reflected analyzing beam is focused on the imaging camera through the camera's objective. Images were acquired exactly as has been done with the schlieren set-up. Although this shadowgraphy set-up provided acceptable spray visualization and determination of key characteristics such as penetration length, it is worth noting the key difference between shadowgraph and schlieren method. First shadowgraph responds to the second derivative of the refraction index while schlieren responds to the first derivative. Hence, shadowgraphy displays the ray displacement from the deflection rather than the deflection angle as is the case of schlieren. This yields in the cases pertain to this work weaker sensitivity of the shadowgraphy. Second, a shadowgraph is as the name suggests merely a shadow image while schlieren is a focused optical image. Finally, the schlieren method requires a cut-off of the refracted light while no such cut-off is needed in the case of Shadowgraph.

4.2 Soot Incandescence and Natural Luminosity Measurement

Hot Soot particles attain the combustion gas temperature in about 10^{-50} – 10^{-6} s within the cylinder and incandesces as *black body*. This fundamental characteristic allows the indirect measurement of the in-cylinder soot concentration by means of acquisition of soot or for that matter flame natural luminosity. The pioneer work of Mueller and martin [119] and Pickett and Siebers [52] of Sandia National Lab have shown the applicability of this approach for the measurement of the in-cylinder soot volume fraction. Acquisition of soot luminosity (incandescence) can be done through imaging and non-imaging light emission acquisition technique. In this work soot or natural luminosity was acquired through both, non-imaging as well as imaging techniques. Practically speaking, the technique is quite straight forward and is based on collecting emitted light or luminosity from the combustion cylinder out to the acquisition device.

The emitted light is certainly broad band and upon integrating it spatially over time, hence summation of light intensity all over the surface section of the acquisition device, one obtains a time-resolved Spatially Integrated Soot Incandescence [SISI] or Spatially Integrated Natural Luminosity [SINL] signal whose peak value is used as a relative measure of the average in-cylinder soot volume fraction. Although it is quite simple and easy to apply approach, careful assessment of the interpretation of the collected emitted light and its association to soot concentration has to be established given the fact that several factors would affect the soot or natural luminosity signal. This is being done next.

Interference from Chemiluminescence in the light of presence of several electronically excited gaseous species or radicals produced during chemical reaction such as water and fuel vapor as well as CO CO2, and OH, CH, C2, HCO radicals has been found to be 4 to 5 order of magnitude smaller than the SISI [120]. In this work a broad and narrow-band natural luminosity measurement (i.e., measurement of light emission without and with spectral filtering) has been performed and yielded less than 3% difference as shown in figure 4-8.

Soot particle properties such as particle size, refractive index and temperature in the light of black body incandescence have certainly immediate effect on the SISI but each of different weight. Considering that Natural Luminosity [NL] from a radiating soot particle with a number density that is uniform over the measurement volume can be expressed in the form as shown in equation 4-3

$$S(\underline{x}) = K f_v(\underline{x}) g(T(\underline{x}), m) \tag{4-3}$$

where $f_v(\underline{x})$ is the soot volume fraction which reflects the weight of particle size and $g(T(\underline{x}), m)$ is function of temperature and refractive index.

Mueller and Martin [119] have shown that even for broad band natural luminosity where the refractive index varies slightly with wave length, the soot or natural luminosity is proportional to T^5 and thus proving that the weight of changes in soot particle or refractive index relative to variation in temperature is certainly negligible and soot incandescence is practically temperature-dependent mainly.

In-cylinder temperature effect that could affect soot incandescence has been kept throughout this work unchanged by maintaining same amount of injected fuel added to the combustion system under all operating conditions under investigation. Thus, changes in SISI signal reflects indeed changes in soot volume fraction.

Spatial distribution of luminous soot could also affect the SISI signal. It has been already shown, however, that given the dependency of soot incandescence on temperature, it is evident that the hottest soot would contribute the highest weight of the signal. The question remains, however, where is the hottest soot located within the combustion chamber? It is a well established knowledge that there are two different regions in the fuel jet where chemical reactions produce significant heat release [120].

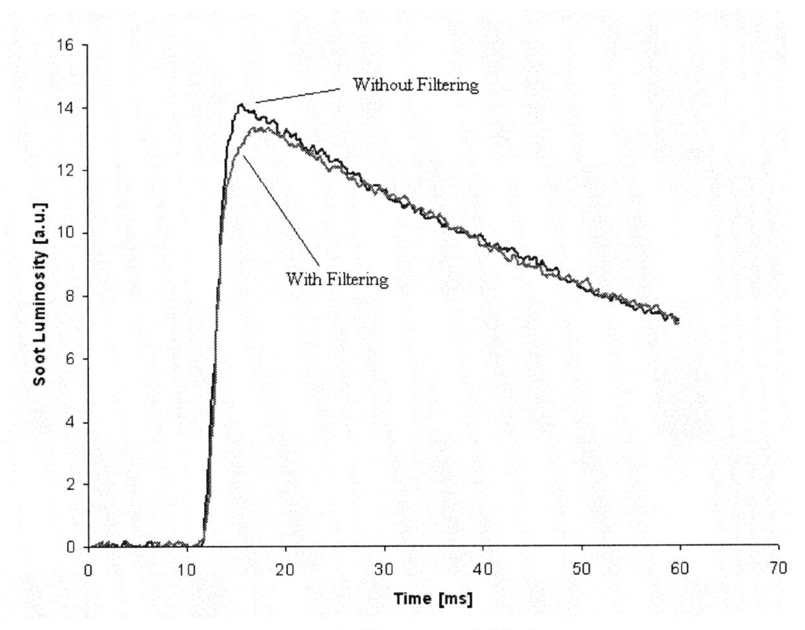

Figure 4-8: SISI signal with and without spectral filtering

The first is the rich, premixed zone downstream near the end of the vaporizing fuel spray while the second region is the diffusion flame that defines the perimeter of the reacting jet. In the later region the mixture is approximately stoichiometric where combustion reaches temperatures significantly higher than that of the former region. The SISI signal in the diffusion flame region, as of a result, could be 2 order of magnitude higher than that of the premixed region. It is expected therefore, that upon the onset of the diffusion flame soot incandescence would originate from the rich side of the diffusion flame. Because of such effect even non-imaging time-resolved SISI could provide information on soot originated upon the onset of diffusion flame in comparison to soot originated from the premixed region.

Furthermore, given the strong dependency of soot incandescence on temperature neither SISI nor SINL should be used to estimate the relative average in-cylinder soot volume fraction late in the expansion stroke. This is mainly because soot during this period may be present but non-luminous as a result of strong cooling effect brought about by expansion process. Mueller and Martin [119] have suggested an applicability limit of this technique up to the time at which the combustion chamber volume has increased by 50%.

The non-imaging soot incandescence acquisition optical set-up is shown in figure 4-9. The light emitted from incandescing soot within the combustion chamber is reflected through the RCM's mirror outward and is then focused through a focusing lens onto the small cross sectional area of the photodiode, maximizing as of a result the intensity of the falling incoming light emission.
Upstream the focusing lens, a band pass filter with the central wave length at 690 nm is been mounted to first filter out radiation from chemiluminescence and second to

maximize the power of the light emitted from soot particles since the emissive power of a black body at this wave length for temperatures typical of diesel flame approaches its maximum (see the spectral black body emissive power in appendix II-B). This also generates a uniform refractive index. A silicon photo diode of the type FDS100 of THORLABS whose sensitivity spectrum is shown in appendix II-A is used to collect the soot luminosity and its output signal is amplified and recorded by a recording computer that employs the labview program. Such signal provides a time-resolved in-cylinder averaged soot luminosity which yields as of a result the temporal development of soot formation. The sensitivity of the photodiode set-up has been found to be insensitive to the room lightening.

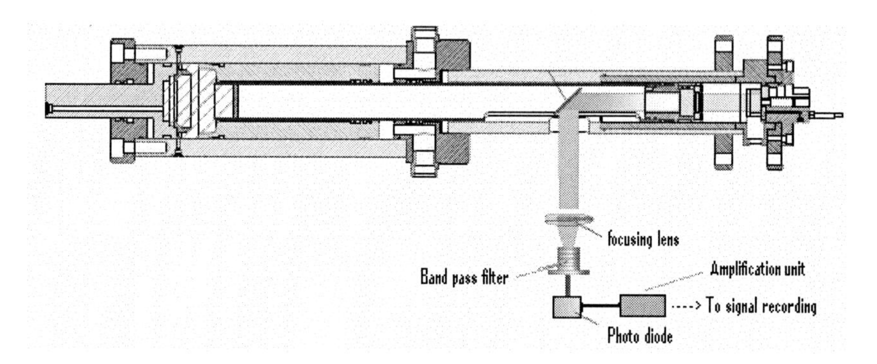

Figure 4-9: Non-imaging photo-diode-based soot incandescence measurement set-up

In the investigations of the effect of injection strategy on the soot formation characteristics of GTL fuel, a second set-up has been deployed which utilizes a High Speed Intensified CCD [HS ICCD] camera to provide images of the soot luminosity for examining the spatial development of soot within the cylinder. This new set-up is shown in figure 4-10

Thus, the photo diode set-up has been replaced by a LaVision HS ICCD camera, type, HSS4 G- HS IRO, with an objective to focus the imaged combustion chamber onto the camera's chip. The band pass filter had to be removed given the very low quantum efficiency and sensitivity of the camera's image intensifier at this specific wave length of 690 nm (see appendix II-C). The gain of the intensifier has been set to very low level at 40% which corresponds to almost neutral intensification while the gating time has been set to 10 μs. Musculus of sandia national lab [124] has applied, however, longer gating time, 35 μs but reduce the camera's aperture. The combination of widely-opened aperture and shorter gating time was found more preferable so to better control signal intensity and avoid damaging the sensitive camera's chip. The acquisition rate has been set to the same rate used to acquire shadow images, hence, 15000 fps. Given that obtaining information on the time-resolved temporal soot formation is also of great importance in these investigations, a special function has been incorporated into the DAVIS processing program of the LaVision camera which allowed the spatial integration of the luminosity signal of each acquired image providing thus a time-resolved SISI.

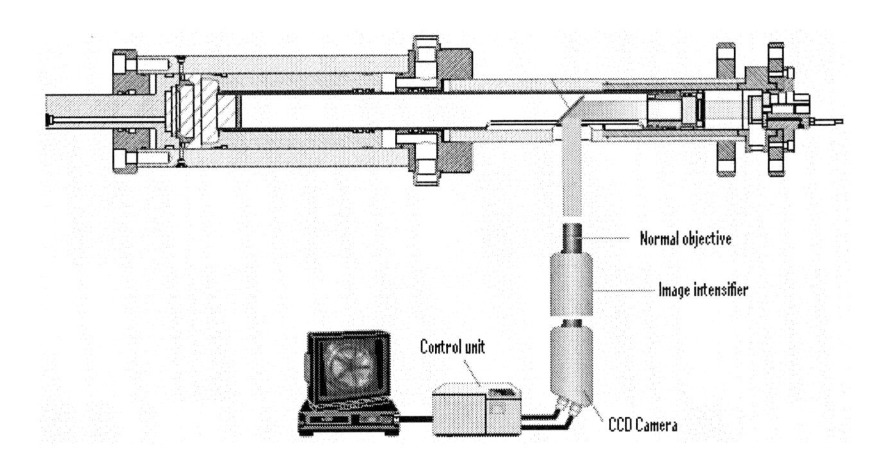

Figure 4-10: Optical set-up of the imaging soot incandescence measurement technique

4.3 OH Chemiluminescence Measurement

Optical emission from excited-state OH radical is a commonly used diagnostic in many fields of combustion research as it provides knowledge on flame-zone location as well as oxidation sites which is substantially important for the understanding of combustion and pollutant formation characteristics of a given system.
It is based on the acquisition of the emitted luminescence from the excited OH [marked hereafter as OH*] which releases photons upon transition from the excited-state back to the basic energy level. Such transition from excited to relaxed level corresponds to a transition in all three energy levels, electronic, vibration and rotation and the intensity of the emitted luminescence at given wave length is directly proportional to the concentration of the excited molecule, *i.e.* OH radicals [5]. Since the excitation energy is the internal energy released as a result of the chemical reactions associated with the combustion process, the emitted luminescence is given the term *Chemiluminescence*.

Dec and Espey [13], among many others, have examined the emission spectrum of diesel combustion and found that diatomic molecule such as OH exhibits an emission spectrum with two peaks. One major peak exhibiting strong intensity, while a second peak exhibiting very weak intensity. The first major peak is in the ultra violet range and is centered around wave length of 307-308 nm. Furthermore, they have indicated that diesel combustion is distinguished by strong emission from CH molecule/radical which seems to dominate the entire flame emission spectrum.
Hall and Peterson [120] have shown as a result that OH* is formed mainly through CH oxidation reaction and the elementary reaction responsible for such formation is;

$$CH + O_2 \leftrightarrow OH^* + CO$$

Measurement of OH chemiluminescence has been deployed in this work mainly to examine the temporal evolution of OH concentration as an indicator of soot oxidation activity prompt by post-injection. The aim is to verify effect of post-injection timing on in-cylinder soot oxidation which reflects soot burn out or destruction.
Such measurement could have been performed by utilization of a photomultiplier on which the emitted luminescence from the combustion chamber is collected following passing through a 308nm band pass filter as done by Aleiferis et.al. [121]. The relatively very week OH chimeluminescence has impeded the application of the photomultiplier because high intensifying factor was necessary. Instead the HS LaVision intensified Camera described in section 4.2 has been used on which an UV objective has been mounted so the combustion chamber is focused on the camera's chip. In order to maximize the signal intensity an OH filter with 50% transmission has been mounted on front of the camera set-up. Simpler OH filter offering around 10% transmission yielded a very week signal and rendered the measurement quality very poor. The camera gain has been set to 50 % while gating to 1 µs. Shorter gating is desired to avoid interference from soot incandescence and 1 µs has been found to be an optimal trade-off yielding strong signal for all investigated cases while simultaneously avoid signal saturation.
The acquisition rate has been set to the same rate of 15000 fps as in the case of soot incandescence and shadowgraphy imaging. The OH-chemiluminescence optical set-up is shown in figure 4-11.

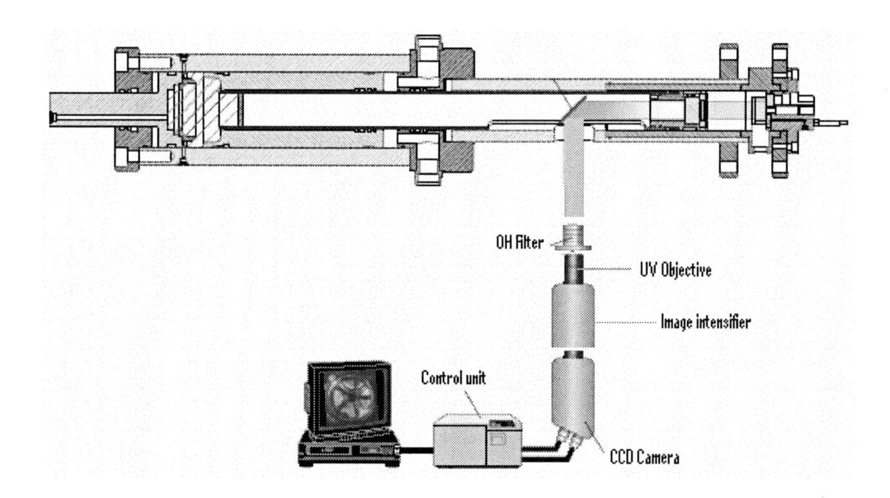

Figure 4-11: OH Chemiluminescence optical set-up

In order to obtain the time-resolved OH chemiluminescence evolution the intensity signal of each image has been spatially integrated applying the same function used for the establishment of time-resolved SISI.

4.4 Flame Self-luminescence Imaging Technique

The flame self-luminescence technique has been applied mainly for the purpose of detecting flame zones aiming determining Lift-Off Length [LOL] (*i.e.* location of flammable sites with respect to the injector nozzle position) which is later correlated to soot formation tendencies.

Like OH chemiluminescence, upon the onset of combustion which is triggered by the associated chemical reactions, photons are released as a result of their transition from an excited-state back to basic energy level.
Generally speaking flame self-luminescence is characterized by broad band emission spectrum. This is mainly, unlike the case of OH chemiluminescence, photons are released from multiple combustion by-products and radicals each emitting light at different wave-length.
Special care should be taken, however, to distinguish flame self-luminescence from soot incandescence. If recalling section 4.2 soot incandescence is multiple orders of magnitude stronger than chemiluminescence. It will overtake, therefore, its emission signal. Imaging, was taken, therefore, only during the very early stage of flame development which provided the necessary information as pertains to LOL and timing of onset of combustion.

The optical set-up is quite simple and is shown in figure 4-12. It consists mainly of a HS CCD camera on whose chip the emitted luminescence is collected. A normal objective has been mounted to focus the combustion chamber on the desired image size. Here again images were acquired at a rate of 15000 fps and exposure time equals to 1/fps. The objective aperture has been reduced in order to attenuate the incoming emission signal and to allow proper visualization of the development of flammable sites.

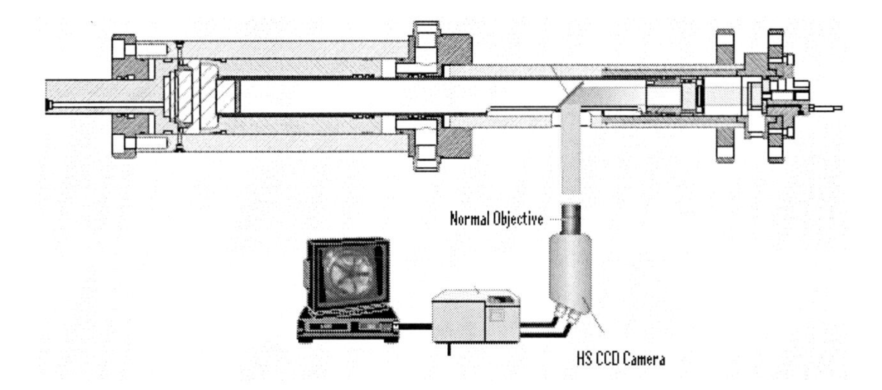

Figure 4-12: Flame self-luminescence optical set-up

5 Physical Parameters Effects on Combustion and Soot Formation of GTL Fuel

5.1 Compression Temperature Effects

Temperature effects on soot formation are of paramount importance as described in the previous sections. On one hand, it affects the soot formation chemistry as well as the mixing and ignition processes and their subsequent combustion development on the other. Reducing cylinder temperature in modern diesel combustion systems has become the main driving mechanism for pollutant reduction. In the practical sense, temperature reduction is being realized through the utilization of EGR. In the case of GTL fuel, however, because of its high Cetane number, this method renders impractical if Compression Ratio [CR] is kept relatively high. The more efficient and practical strategy that will result from this goal is to lower the CR. The ultimate objective is to arrive at a combustion mode in which the injection process is completed prior to the onset of combustion. In other words, fuel injection must be completed within ignition delay time. This guarantees that the injected fuel burns in a premixed mode which promises reduction in soot formation to levels capable of meeting future emissions regulations. As previously mentioned, the combination of lower CR and GTL fuel allows the realization of this goal.

To qualitatively demonstrate this, three cases were studied in which cylinder temperature was varied through varying CR, namely CR=16,15 and 14 while injection pressure, injection timing, initial cylinder pressure and cylinder wall temperature remained unchanged. The critical factors were cylinder pressure and temperature at the time of the Start Of Injection [SOI]. The cylinder pressure at SOI was determined from the measured time-resolved cylinder pressure using the Kistler pressure transducer installed in a cylinder head. However, temperature assessment had to be calculated using the thermodynamic relations, namely the entropy equation and ideal gas law as follows:

The entropy equation;

$$ds = \frac{1}{T}dh - \frac{1}{T}vdp$$

under the assumption of thermally perfect gas (hence $h = h(T)$ only) the entropy equation becomes

$$ds = c_p\frac{dT}{T} - v\frac{dp}{T}.$$

Using the ideal gas law; $pv = RT$ whih can be rewritten as $\frac{v}{T} = \frac{R}{p}$ the entropy equation can be rewritten as

$$ds = c_p\frac{dT}{T} - R\frac{dp}{p}$$

Assuming calorically perfect gas where $c_p = constant$ and isentropic process, hence s = constant we obtain :

$$\frac{P_2}{P_1} = \left(\frac{T_2}{T_1}\right)^{\frac{\gamma}{\gamma-1}}, \tag{5-1}$$

where: $\frac{\gamma}{\gamma-1} = \frac{c_p}{R}$

Thus, knowing the cylinder pressure at SOI and using equation 5-1, the cylinder temperature can be calculated. The measured cylinder pressure and calculated temperature for the three investigated cases are tabulated below

CR	P @ SOI [Bar]	T @SOI [K]
14	32.8	813
15	35.65	830
16	38.45	847

To better evaluate temperature effect on soot formation, a qualitative assessment of the combustion characteristics of the three investigated cases in terms of heat release analysis has been performed. This has been carried out through the deployment of a simple heat release model based on the measured time-resolved cylinder pressure and volume. This simple model was derived from the first law of thermodynamics while first taking into account that the *net heat-release,* which is the difference between the apparent *gross heat-release* rate dQ_{ch}/dt (*i.e.* heat release by combustion of the fuel) and the heat-transfer rate to the walls dQ_{ht}/dt, equals the rate at which work is done on the piston plus the rate of change of sensible internal energy of the cylinder contents, and second assuming that the cylinder content can be modeled as ideal gas as follows:

$$\frac{dQ_n}{dt} = p\frac{dV}{dt} + mc_v\frac{dT}{dt} \tag{5-2}$$

Using the ideal gas law, $pV = mRT$, we obtain

$$mdT = \frac{1}{R}(PdV + VdP) \tag{5-3}$$

Substituting equation 5-3 into equation 5-2 we obtain

$$\frac{dQ_n}{dt} = \left(1 + \frac{c_v}{R}\right)p\frac{dV}{dt} + \frac{c_v}{R}V\frac{dP}{dt} \tag{5-4}$$

Using the relation $\frac{c_v}{R} = \frac{1}{\gamma - 1}$, equation 5-4 can be rewritten as

$$\frac{dQ_n}{dt} = \frac{\gamma}{\gamma - 1} p \frac{dV}{dt} + \frac{1}{\gamma - 1} V \frac{dP}{dt} \qquad (5\text{-}5)$$

Where γ has been assumed constant and of value of 1.34.

Using this heat release model (*i.e.* equation 5-5) together with the measured time-resolved cylinder pressure and volume, time-resolved curves for cylinder pressure and Rate of Heat Release [ROHR] were established for the three different investigated cases and are shown in figure 5-1.

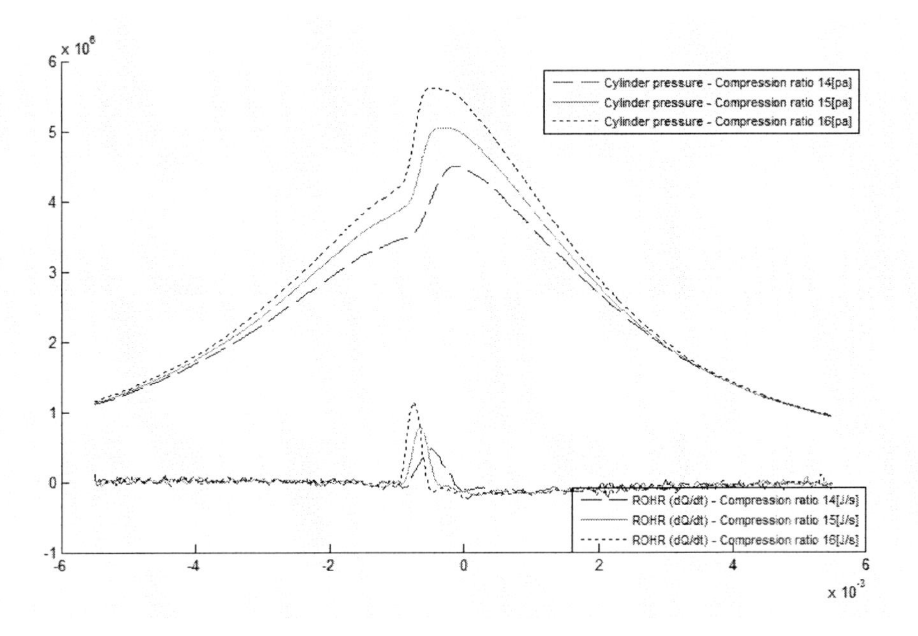

Figure 5-1: Cylinder pressure traces and time-resolved Rate Of Heat Release for the compression ratio cases, CR=14, 15 16.

The immediate observations depicted in figure 5-1 are first that GTL combustion exhibits fast combustion characteristics where the deep dip typically seen in pressure traces of conventional diesel following the end of compression and prior to initial pressure rise is not seen even in the case of the lowest CR=14. This can be attributed at large to the high cetane number and relatively low distillation temperatures of GTL fuel. Second, is that the initial rise in ROHR is significantly delayed for the case of lowest cylinder temperature (CR=14) which indicates a prolonged ignition delay. The gain in prolonging ignition delay for the purpose mentioned earlier is about 500 μs. This is typical injection duration at medium load.

Thus, such significant time increase allows the completion of fuel injection prior to the onset of combustion, which would have a direct impact on the soot formation as would be shown later.
This observation shall be revisited in later section where spray visualization shall be deployed to further elaborated it

As a result of the prolonged ignition delay and late rise in heat release, the peak cylinder pressure is shifted closer to Top Dead Center [TDC] towards the end of the compression stroke yielding a shorter time period before expansion recovers and relaxes cylinder pressure and temperature which ultimately reduces soot level.

Furthermore, the lowest cylinder temperature realized at CR=14 yielded the least steep slope in heat release rate and cylinder pressure rise bringing combustion noise to lowest levels.

The in-cylinder soot formation of these three cases was assessed by applying the non-imaging time-resolved Spatially Integrated Soot Incandescence [SISI] measurement technique which provided a measurement of the time-resolved cylinder-averaged soot volume fraction as described in the previous chapters. This is very useful for comparative analysis purposes. The results are shown in figure 5-2 from which one can draw the following observations:

1. The lowest cylinder temperature deduced by CR=14 delays the first appearance of soot by a time period comparable to the delay in heat release. This is in agreement with the fact that soot luminosity appears slightly after initial flame luminosity.
2. The lower the cylinder temperature is, the lower the soot formation rate which relates to the lower rate in heat release shown earlier.
3. The lowest cylinder temperature yields the lowest peak in soot concentration. This demonstrates the strong dependency of soot formation on temperature on one hand while on residence time available for soot formation on the other. Because for lower temperatures, the peak in cylinder pressure shifts towards TDC. Thus, the residence time available for soot formation before expansion takes place which leads to a temperature drop is shorter.
4. The rate of soot oxidation taking place in the expansion stroke (in the early stage of this process during which this measurement technique is valid) does not seem to be strongly sensitive to a change in CR. This is mainly because the oxygen content, which is the prime trigger of soot oxidation, is quite comparable in all cases and in none of the case additional heat is added towards start of expansion stroke.

It is worth recalling the effect of temperature on the Lift-Off Length [LOL] in order to explain why soot formation increases as cylinder temperature increases. Because LOL decreases as temperature increases, the combustion zone shifts closer to the injector nozzle where mixture zones are fuel-richer leading to a higher formation rate of soot or soot precursors. However, the critical factors here are the portion of fuel that burns in a diffusion flame and the residence time available for soot formation.

Due to high Cetane number of a GTL fuel, lowering the cylinder temperature by lowering CR seems to be, indeed, the preferable method.
This is because it one hand prolongs the ignition delay to time period that allows completion of fuel injection prior to the onset of combustion reducing, as a result, the amount of fuel that burns in a diffusion flame while it provides on the other the necessary chemical composition in terms of the air-fuel ratio, which in the light of the specific fuel properties of the GTL fuel allows fast combustion, yielding a shorter residence time during which soot can form. Thus, in the case of GTL fuel, lowering CR as low as 14 allows the realization of a combustion mode that is referred to in this work as Low CR Premixed [LCRP] Combustion.

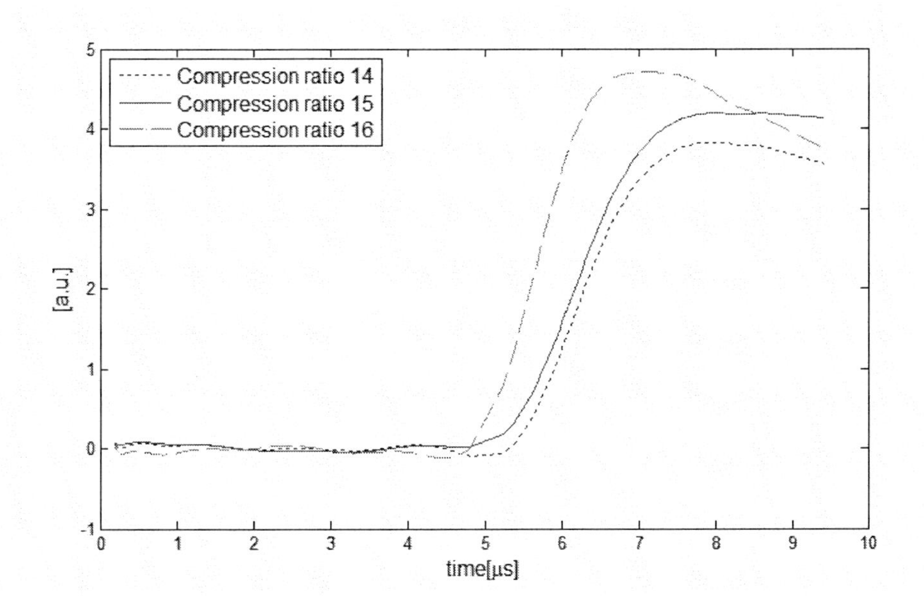

Figure 5-2: Time-resolved Spatially-Integrated Soot Incandescence for the compression ratio cases, CR=14, 15, 16.

Furthermore, reduction in CR from 16 to 14 and thus reduction in the cylinder temperature at SOI from 847°K to 813°K yields more than 30% reduction in soot formation peak.
This end result is the outcome of the simultaneous reduction in both compression temperature as well as compression pressure brought about by reduction in CR. In the next section, the compression pressure effect is decoupled from temperature effect and its influence on soot formation will be demonstrated.

5.2 Compression Pressure Effects

To investigate the effect of compression pressure on soot formation independent of temperature, the CR was fixed at 14 and injection pressure, duration and timing were kept unchanged while varying charge (boost) pressure only. Thus, three cases of P_{CH} =1.05, 1.3 and 1.5 [bar] yielding cylinder pressure at SOI of P_{SOI}= 32.8, 40.66 and 46.88 [bar] respectively were investigated. To realize this, the driving pressure of the RCM had to be adjusted in accordance with the charge pressure to assure that the cylinder piston reaches the desired end piston position corresponding to the fixed CR deployed and, thus, guaranteeing that cylinder compression temperature remains unchanged. To assure this, the ratio of $\frac{P_{SOI}}{P_{CH}}$ in equation 5-1 has been verified to be almost unchanged for the three investigated cases. This ratio has been found to be 31.24, 31.27 and 31.25 for P_{CH} =1.05, 1.3 and 1.5 respectively. The negligible inequality can be attributed to the cyclic variation of the RCM.

Here as well, the non-imaging time-resolved SISI technique was applied to measure the in-cylinder averaged soot volume fraction. Figure 5-3 depicts the time-resolved soot incandescence (luminosity) for three studied cases.

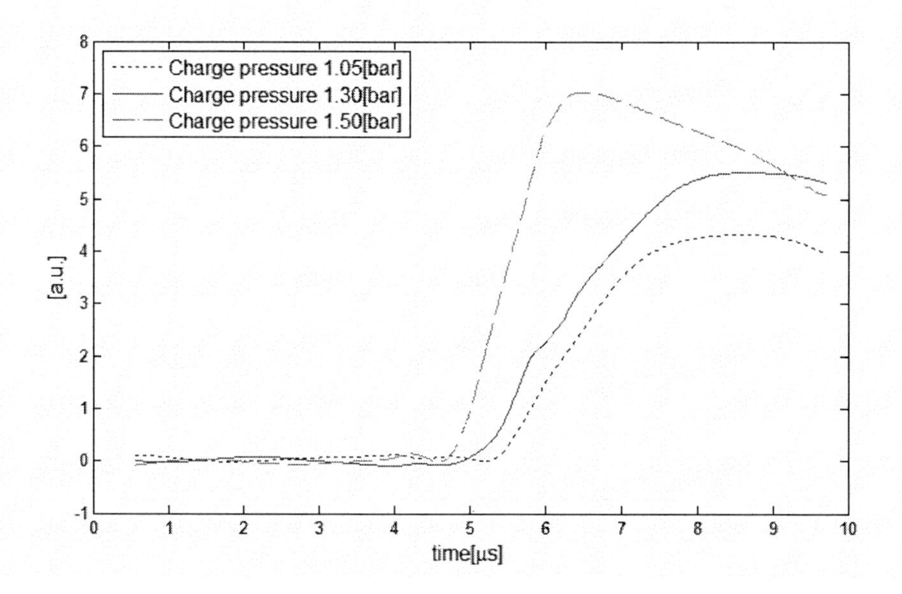

Figure (5-3): Time-resolved Spatially-Integrated Soot Incandescence for the compression ratio cases, P_{CH} =1.05, 1.3, 1.5.

Figure 5-3 shows that the peak in soot formation increases non-linearly with an increasing cylinder gas pressure (hence density) which is in agreement with the work of Pickett and Siebers [52]. This recalls the influence of Lift-Off Length on soot formation and the dependency of the former on cylinder gas pressure (density).
To further elaborate such dependency, imaging of flame luminosity prior to the onset of soot has been taken using the high speed CCD camera in order to obtain a qualitative assessment of the lift-off length. A representative images showing flame location in a single fuel jet with respect to injector's nozzle position for the three investigated cases are shown in figure 5-4. An image post processing program capable of reading signal intensity has been utilized to measure the lift-off length. The flame zone has been determined by a transition from zero to high intensity reading along the jet axis.

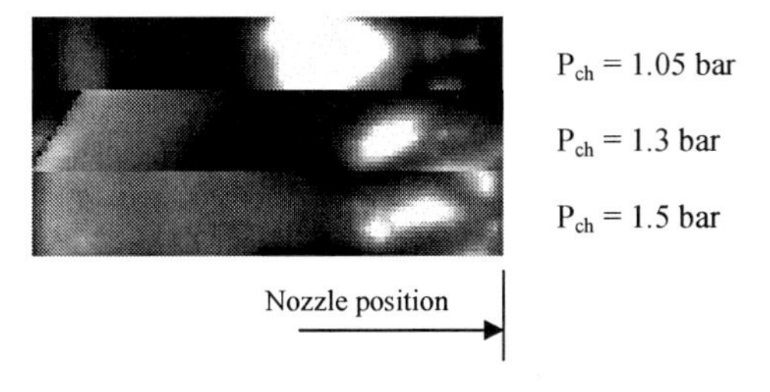

Figure 5-4: Flame location in a single jet with respect to nozzle position for three charge pressure cases for LOL measurement

The measured LOL has been plotted vs. charge pressure and is shown in figure 5-5. A negative quadratic relation seems to govern the relation between LOL and cylinder charge pressure or effectively compression pressure by which the LOL scales with the square of charge pressure.

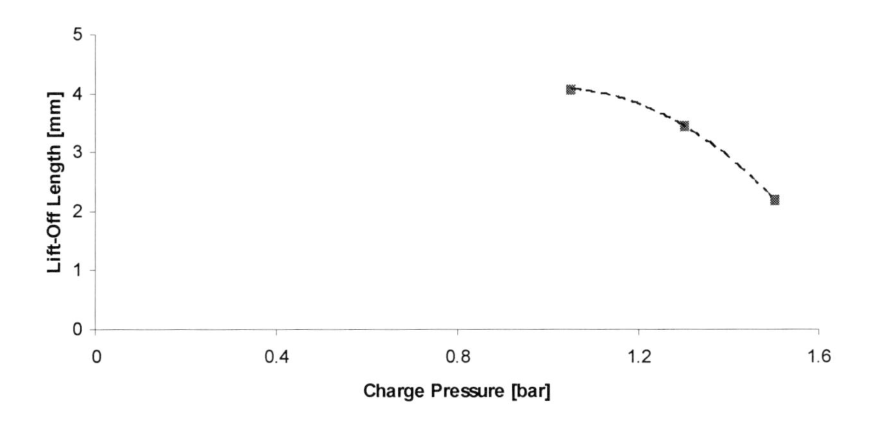

Figure 5-5: LOL as function of charge pressure

Recalling the work of Mueller and Wittig [86], whose work related to same dependency of soot on gas pressure, indicates, first, that LOL has the strongest influence on soot formation and, second, that compression pressure (or cylinder charge density) effect on soot formation of GTL fuel seems to have a similar trend like conventional diesel fuel.

Further trends that are observed in figure 5-3 are that soot appears earlier as gas pressure increases. This can mainly be attributed to the effect of gas pressure in first shortening ignition delay time and, second, in intensifying fuel jet break-up as a result of an increasing gas density. The rate of soot production increases as gas pressure also increases. This is mainly due to increase in the rate of the soot formation chemical reactions and transport with increasing gas pressure.

Soot oxidation, on the other hand, seems to be stronger for higher gas pressures. This is clear from the decaying slopes of the soot luminosity signals whose decay is the strongest for the highest charge pressure case. This is mainly because a higher gas pressure increases the oxygen content and for same amount of injected fuel mass, the higher the oxygen available in the post combustion phase, the stronger the soot oxidation effect is. Gas pressure or oxygen relative pressure seems, however, to have a non-linear effect.
Nevertheless, assessing the final soot level, within the validity of this technique during the expansion stroke, shows that an increase in cylinder gas pressure does not contribute physically to a tangible reduction in final soot production. In order to enhance soot oxidation, cylinder pressure has to be boosted to relatively high levels; and despite the stronger soot oxidation associated with such pressure elevation the benefits brought about for reducing final soot level are quite minimal, if at all. The final soot level is the lowest for the lowest charge pressure which corresponds to a pressure slightly higher than ambient pressure (hence natural aspiration).

Initial cylinder charge pressure will be set, therefore, at this lowest pressure in all subsequent investigations in this work.

6 Injection Parameters Effects on Combustion and Soot Formation of GTL Fuel

6.1 Injection Pressure Effects

To investigate the effect of injection pressure on soot formation, the non-imaging time-resolved SISI technique has been applied for three different injection pressures, namely $P_{INJ.}$=1050, 1350 and 1600 [bar] while keeping injection timing unchanged. Special care was taken to assure that the injected fuel mass is the same for all cases by adjusting injection duration in accordance with the following relation that relates injected fuel mass to injection pressure and injection duration:

$$m_f = C_D(A_n / m^2)\sqrt{2(\rho_f / Kg / m^3)(\Delta P_{INJ} / Pascals)}(\Delta t / s) \qquad (6\text{-}1)$$

where C_D and A_n are orifice flow rate coefficient and total orifice area respectively.

The measured soot incandescence (luminosity) for the investigated three pressures is shown in figure 6-1 which shows that under the condition of low CR, namely 14, injection pressure does not seem to have an effect on the inception (first appearance) of soot. This can be attributed to the fact that initial soot formation is strongly temperature-dependent and for low aromatic fuel this becomes of paramount importance.

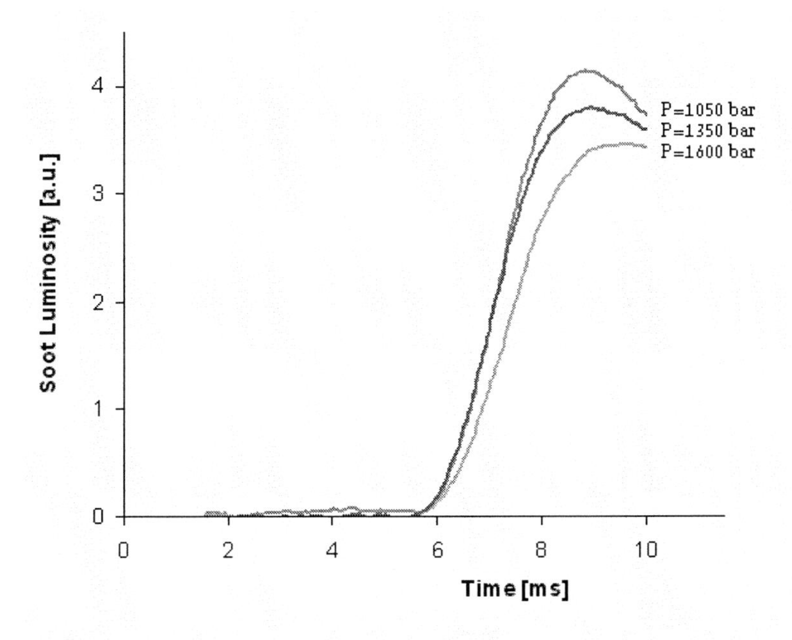

Figure 6-1: Time-resolved Spatially-Integrated Soot Incandescence for the injection pressure cases, $P_{INJ=}$ 1050, 1350, and 1600 [bar]

The effect of injection pressure is sensible, however, in its effect on the formation rate and soot peak. As seen in figure 6-1, the rate of soot production (the slope of the rising soot luminosity curve) decreases with increasing injection pressure. Such a decrease is more tangible at relatively high injection pressure, the case of P_{INJ}=1600 [bar] namely. This is attributed to the effect of higher injection pressure on enhancing fuel atomization resulting in fuel drops of a smaller diameter. The smaller the drop diameter, the faster heat transport is and, thus, the stronger soot oxidizes and the lesser soot agglomerates. The fact that injection pressure has to be significantly elevated for enhancement in atomization to be sensible can be attributed to the lower density and, thus, compressibility of the GTL fuel. Recalling *Weber-Number* given in equation 2-1, one notes that because it scales linearly with density and injection pressure (hence square of jet velocity), if density decreases, injection pressure has to be significantly larger so the resulting magnitude of *Weber Number* becomes sensibly larger. Furthermore, given that the temperature effect on soot production is multiple orders of magnitude greater than the effect of droplet size, under the condition of low CR the fuel drop diameter has to be significantly smaller for a reduction in soot production to be tangible. This, in return, requires that injection pressure is elevated to a relatively high pressure. This emphasizes the advantage of deployment of a high-pressure injection system for the purpose of reducing soot production especially for the low CR combustion system

The effect of injection pressure on a soot production peak is quite obvious. Soot peak decreases with increasing injection pressure. The interesting observation is, however, that soot peak decreases linearly with increasing pressure as shown in figure 6-2. This recalls the effect of injection pressure (hence, the square of jet velocity) on Lift-Off Length [LOL]. Therefore, an assessment of the Lift-Off Length has been performed as done in section 5.2. Figure 6-2 also shows the measured LOL for the three investigated injection pressure cases which clearly shows the linear dependency of LOL on injection pressure which is in agreement with the work of Siebers and Higgins [53]. This elaborates, once again, the strong dependency of soot production on LOL according to which GTL fuel seem to exhibits similar relations as those of conventional diesel fuel.

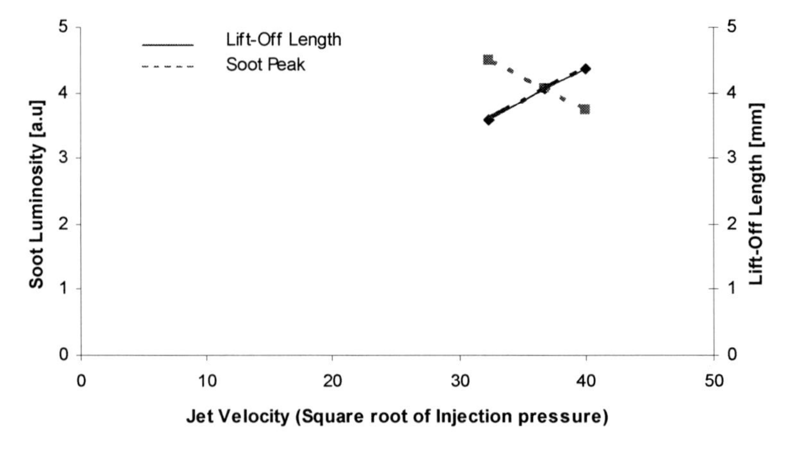

Figure 6-2: Lift-Off Lenght and soot peak as function of jet velocity

The decrease in soot peak in response to increase in injection pressure can be attributed to two more reasons. First is the effect of residence time available for soot formation. By increasing injection pressure the time available for a fluid element to move through the soot-forming region of the fuel jet decreases resulting in lesser time available for soot formation before the oxidation-dominated region is reached.
Second reason is the effect of air-fuel mixing. By increasing injection pressure the velocity of the fuel jet also increases. Thus the fluid penetration into the combustion chamber and its entrainment in air occur also faster providing as a result longer time for air-fuel mixing. In conventional combustion system with conventional diesel fuel, this could lead, however, to an early appearance of soot. This does not happen in the case of combined low CR and GTL fuel as has already been demonstrated which indicates that under these conditions temperature effect seems to overtake the air-fuel mixing effect so increase in injection pressure does not affect start of soot appearance. Rather, it contributes only to the enlargement of the lean mixture zone which bears low propensity to produce soot. In other words, increase in injection pressure increases the soot-less heat-release phase.

6.2 Injection Timing Effects

Effects of injection timing on soot formation have been investigated at constant injection pressure and duration. The three injection timing cases were set in a manner that the resulting peak cylinder pressure occurs Before Top Dead Center [BTDC], at Top Dead Center [TDC] and After Top Dead Center [ATDC] corresponding to SOI timing at 12, 9 and 6 CA BTDC. The pressure traces of these three cases are shown in figure 6-3 from which one can draw a very interesting observation that here again GTL fuel, for all injection timing cases, does not exhibit the deep dip in pressure traces following end of compression typically seen in pressure development of a conventional diesel fuel. This is attributed to the relatively high cetance number and lower T90 distillation temperature. The later exhibits the capability of GTL fuel to evaporate faster which accelerates air-fuel mixing, and combined with a high CN the ignition delay is significantly shortened. Thus, under condition of low CR, cetane number and distillation temperature effects seem to take over the effect of small temperature variation associated with varying injection timing on ignition delay.

Here, as well, the non-imaging time-resolved SISI technique has been deployed to evaluate the effect of injection timing on soot production. Results are shown in figure 6-4 which shows that retarding injection timing slightly decreases soot production rate while, on the other hand, significantly decreases soot production peak.
In order to better explain this behavior, the measured time-resolved cylinder pressure and volume were used to calculate Rate Of Heat Release [ROHR] using the simple heat release model presented in section 5-1.
The results were plotted and shown in figure 6-5. Also shown in this figure is the time at which the onset of soot production of each investigated case occurs (marked with circle on each corresponding curve).

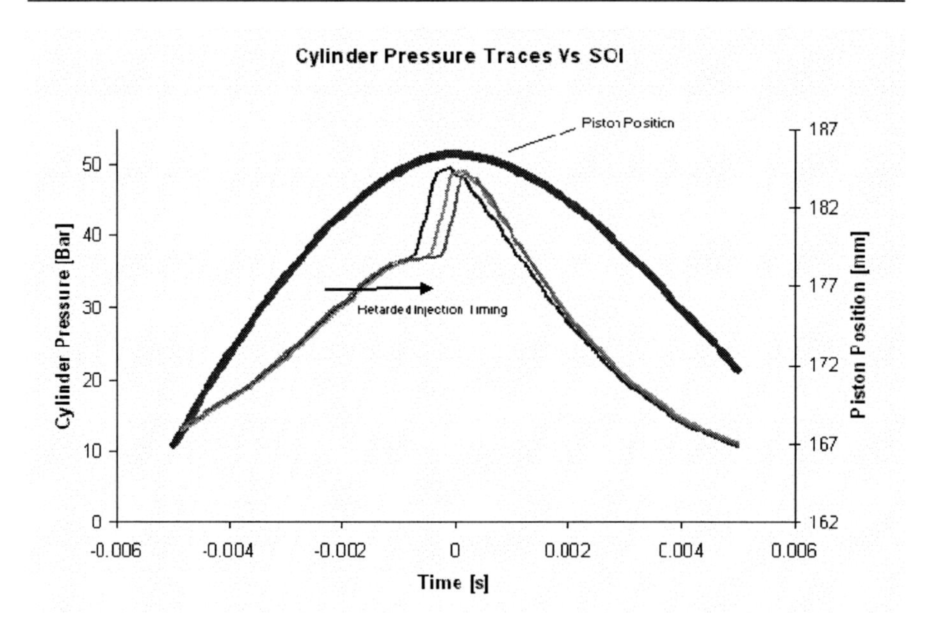

Figure 6-3: Cylinder pressure traces of the three injection timing cases, BTDC, TDC, ATDC with respect to piston position

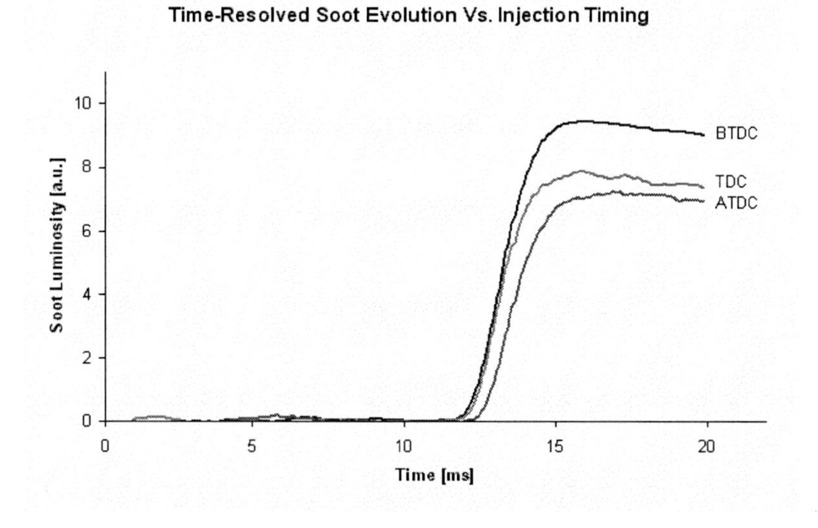

Figure 6-4: Time-resolved Spatially-Integrated Soot Incandescence for the injection timing cases, BTDC, TDC, ATDC

As injection timing is retarded, the portion of heat released before the onset of soot is larger, indicating that larger portion of combustion is soot-less, hence the portion of non-sooting pre-mixed combustion is larger. Such effect of retarded injection timing has also been observed by Fang et. al.[125] in the case of conventional diesel fuel. This emphasizes the dependency of soot formation on temperature and residence time. As injection timing is retarded, the time available for temperature build-up following the cooling effect brought about by injection itself and the time available for soot precursors to form or to further develop into soot is shorter. The concentration of produced soot is, therefore, lower.

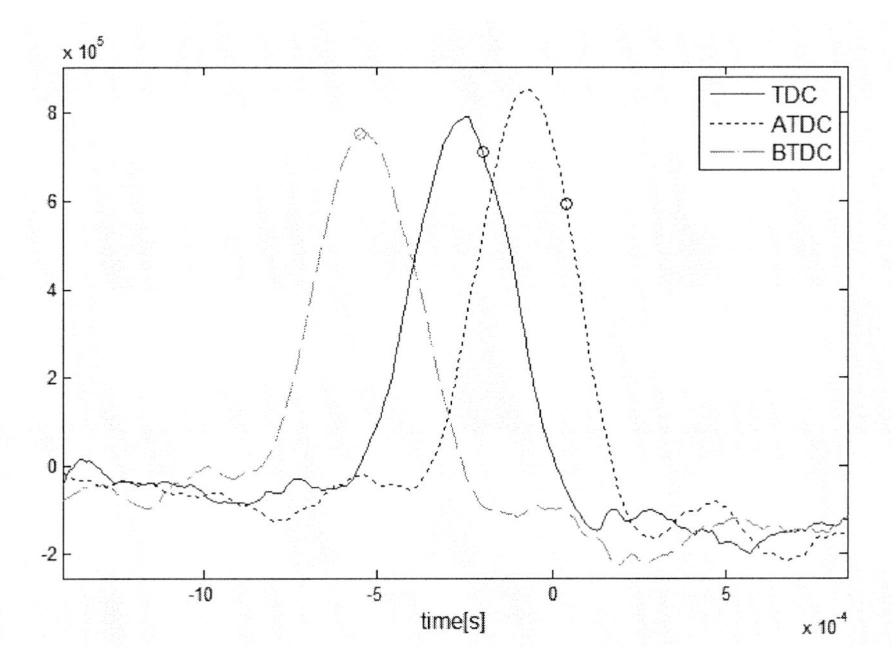

Figure 6-5: Rate of Heat Release [ROHR] curves of the injection timing cases, BTDC, TDC, ATDC. Also shown is the marks (circle) of the onset of soot on each corresponding case

Interestingly, however, is that by retarding injection timing, the ATDC case in particular, soot appears much later after heat release rate has reached its peak which indicates a premixed HCCI-like combustion.

Furthermore, by advancing injection timing, the portion of heat released during compression is larger yielding a higher cylinder temperature and pressure (note the higher cylinder pressure peak for advanced injection timing cases depicted in figure 6-3 which increase and facilitate soot formation.

Retarding injection timing, on the other hand, increases the portion of heat released towards the expansion stroke. Thus, heat addition during expansion and in the light of better mixing due to enhanced flow motion brought about by piston motion during reversing movement direction, soot oxidation increases bringing the final soot to a lower level.

It is also worth noting, that under the operating conditions deployed in this work namely low CR, late injection or retarded injection timing strengthens the effect of cylinder charge cooling brought about by heat lost from the cylinder charge to fuel evaporation.
This effect becomes stronger in case of late injection because in advance or early injection, charge cooling is somewhat balanced by compression. Thus the overall effect on lowering charge temperature is weaker.
Such cooling effect allows, in return, better air-fuel mixing and increases the lean premixed portion of the mixture even to a level of flame-less heat release (combustion). This, in return, lowers the propensity to produce soot. This is a very interesting effect and results suggest that the combination of GTL fuel, low CR and late injection together with sufficiently high injection pressure is a promising strategy for the realization of two advance combustion modes, the so called Highly Premixed Late Injection [HPLI] and HCCI-like strategies even without EGR.

7 Injection Strategy Effects on Combustion and Soot Formation of GTL Fuel

Following the investigation of injection parameters' effect on soot formation characteristics and the establishment of the understanding of injection pressure and timing effect on soot formation, subsequent investigations have been carried out to analyze the influence and merits of an injection strategy on combustion and soot formation of GTL fuel. Injection strategy has become a key aspect of a modern diesel combustion system which is deployed as a measure or a technique to balance the trade-off between combustion efficiency and pollutant emissions. This, in turn, has a direct effect on fuel introduction into the combustion system reflected in the characteristics of spray development on one hand and on combustion and heat release pattern and subsequently on soot formation on the other. The investigations in this part of the work have, therefore, included these three aspects. A Shadowgraphy technique has been used for the visualization of spray development. Time-resolved in-cylinder pressure and volume have been measured for the combustion and heat release analysis while imaging-based soot incandescence measurement has been conducted for the assessment of temporal and spatial soot development.

Three strategies have been investigated in this work, namely Single Main Injection [SMI], Split Injection [SI] compromising of Pilot and Main Injection, and Multiple Injection [MI] compromising of pilot, main and post injection. All deploy constant injection pressure. The working injection pressure had to be lowered, however, to a pressure lower than those investigated in section 6.1, namely to pressure of P_{INJ}=750 bar. This is mainly because the intensity of the light source available for this work was not particularly high and in order to be able to visualize fuel penetration, fuel spray had to be somewhat denser so the relative density gradient between fuel and surrounding is somewhat detectable. As a result, the injection pressure has been reduced to such a lower level.

7.1 Single Injection Strategy Effects

The first case study investigated in this work is *single injection strategy*. This has been deployed at an injection timing referred to as TDC in section 6.2 and injection duration of 0.5 ms. Shadowgraph imaging has been performed to provide qualitative assessment of the spray development and to determine initial ignition sites. Images have been acquired at a rate of 15000 fps and frame resolution of 256x256 pixels. These are shown in figure 7-1. The images only show the left half of the combustion chamber because the positioning of the light source, which had to be somewhat off-axis as already explained, yielded higher reflection from the right side of the combustion chamber which is very much undesired. Furthermore, the type of injector nozzle used throughout this work exhibits a symmetric spray pattern. So as far as spray evolution is concerned, considering half of the view field does not lessen the quality of the information obtained.

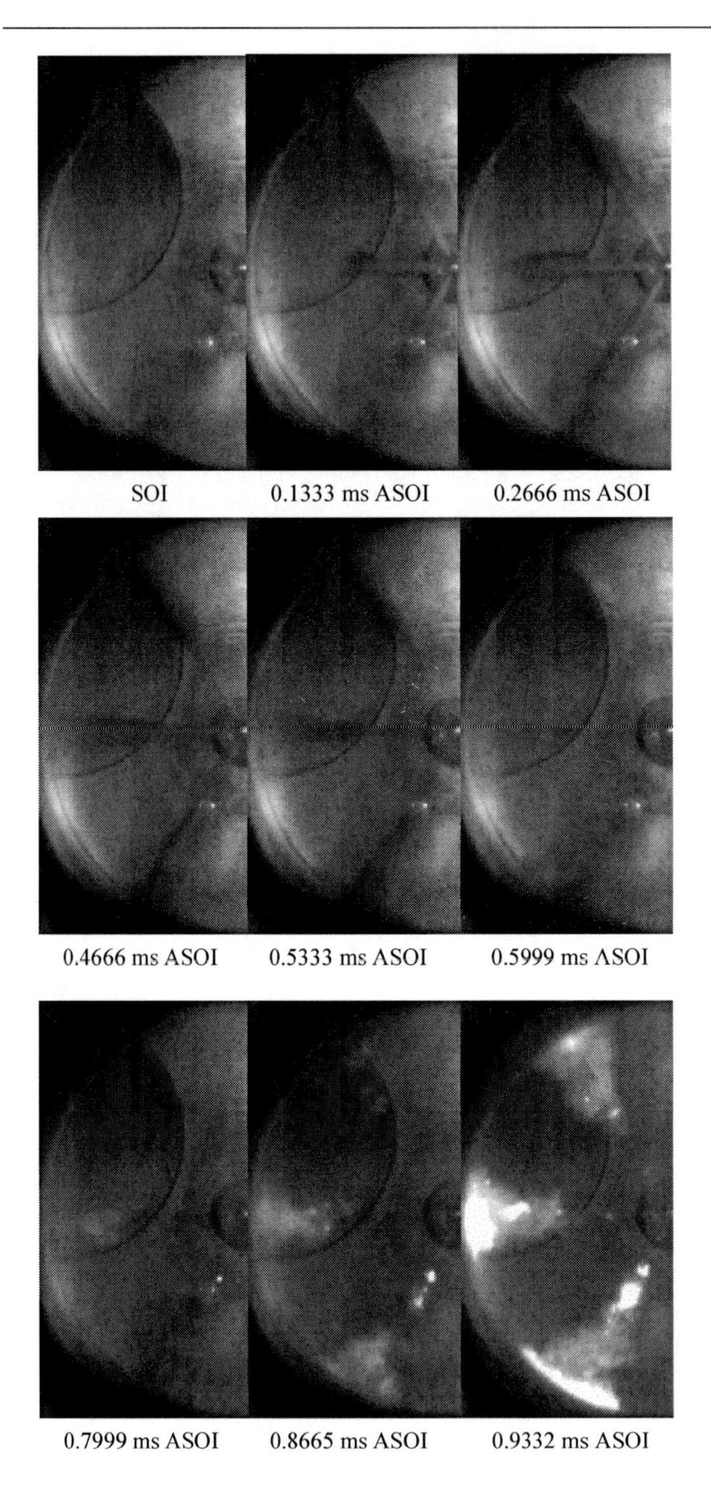

Figure 7-1:
Shadowgraph images of spray development of a SMI

As seen in figure 7-1, the fuel jet, following exiting the nozzle, penetrates monotonically and traverses the entire combustion chamber. (Qualitative assessment of spray penetration length and jet velocities are obtained from these shadowgraph images but shall be revisited in the subsequent subsection). It evaporates completely long before ignition takes place indicating, once again, that lowering Compression Ratio [CR] to CR=14, enables the realization of premixed combustion with GTL fuel despite its high cetane number. As soon as the mixture reaches combustible conditions, ignition seems to take place almost simultaneously downstream as well as upstream, where sufficient air-entrainment has already brought air-fuel mixing to a combustible mixture. Following the onset of such multiple ignition sites the flame grows and spreads relatively quickly, yielding a relatively fast combustion mode. To further elaborate this last observation, calculation of the Rate Of Heat Release [ROHR] based on the simple heat release model presented in section 5.1 has been performed whose results are plotted in figure 7-2 alongside a time-resolved cylinder pressure. The ROHR curve exhibits two distinct stages with a very short transition time separating them. Considering the images acquired at the onset of initial flame luminosity, one can conclude that the first stage beginning at the start of heat release and ending at the onset of flame luminosity marked with blue circle on the heat release curve is the stage of flameless combustion during which cylinder pressure experiences mild increase rate and as a result no soot should be expected to form. The second stage, however, which begins at the onset of flame luminosity, is a combustion of premixed flame which generates a steep increase rate in cylinder pressure and heat release. Such steep rising slop and narrow width of the ROHR and cylinder pressure curves are indeed two strong markers of fast combustion. This in return yields lower production of soot.

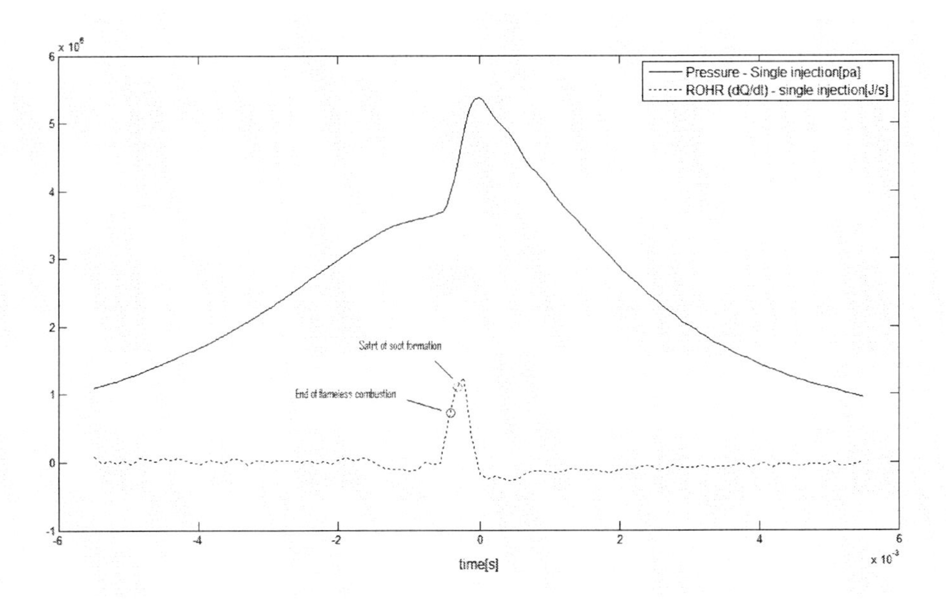

Figure 7-2: Cylinder pressure and rate of heat release curves for SMI. Also shown are the marks of end of flame-less heat release (blue circle) and start of soot appearance (green circle)

To assess the soot formation characteristics of single injection strategy, an acquisition of images of soot luminosity/incandescence has been performed using the HS ICCD camera set-up described in section 4.2.1. at an acquisition rate similar to that of the shadowgraph imaging, hence 15000 fps. This provides time-resolved spatial development of soot within the cylinder. A special function added to the La-Vision software interface allowed the integration of the spatial signal of every acquired image and, thus, providing a time-resolved Spatially Integrated Soot Incandescence [SISI]. This has been used to obtain the general and global soot formation characteristics. Figure 7-3 shows the time-resolved SISI which, in conjunction with the heat release, curve shows that soot appears first only following the onset of premixed combustion (marked in green circle on the ROHR curve) which appears about 110μs after first appearance of flame luminosity while no soot seems to form during the flameless heat release. The SISL also exhibits a narrow width and relatively low peak. This indicates that because GTL under these conditions and following the flameless heat release stage burns in a fast combustion mode, the residence time available for soot formation is relatively short yielding an overall low soot level. Furthermore, because GTL fuel combustion takes place mainly in premixed mode, the level of soot formed is indeed very low.

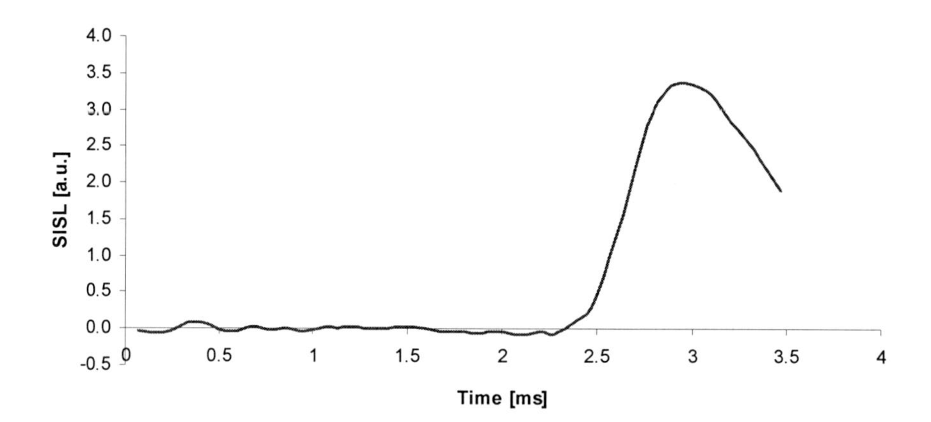

Figure 7-3: Time-resolved Spatially Integrated Soot Incandescence of SMI

To further examine this, one needs to examine the spatial development of soot. We consider, therefore, the soot images acquired with the HS ICCD camera. A set of such images are shown in figure 7-4. By observing the images, one can clearly note the relatively weak luminosity as a result of a reduced amount of soot being formed.
The first soot locations appear at the same locations of those of the initial flame (shadowgraph images). They appear simultaneously up- and downstream referring to two distinct combustion zones. One is the downstream, at the outer spray tip where air entrainment is the stronger and second upstream, at the jets central axis close to the nozzle in the vicinity of the Lift-Off Length.

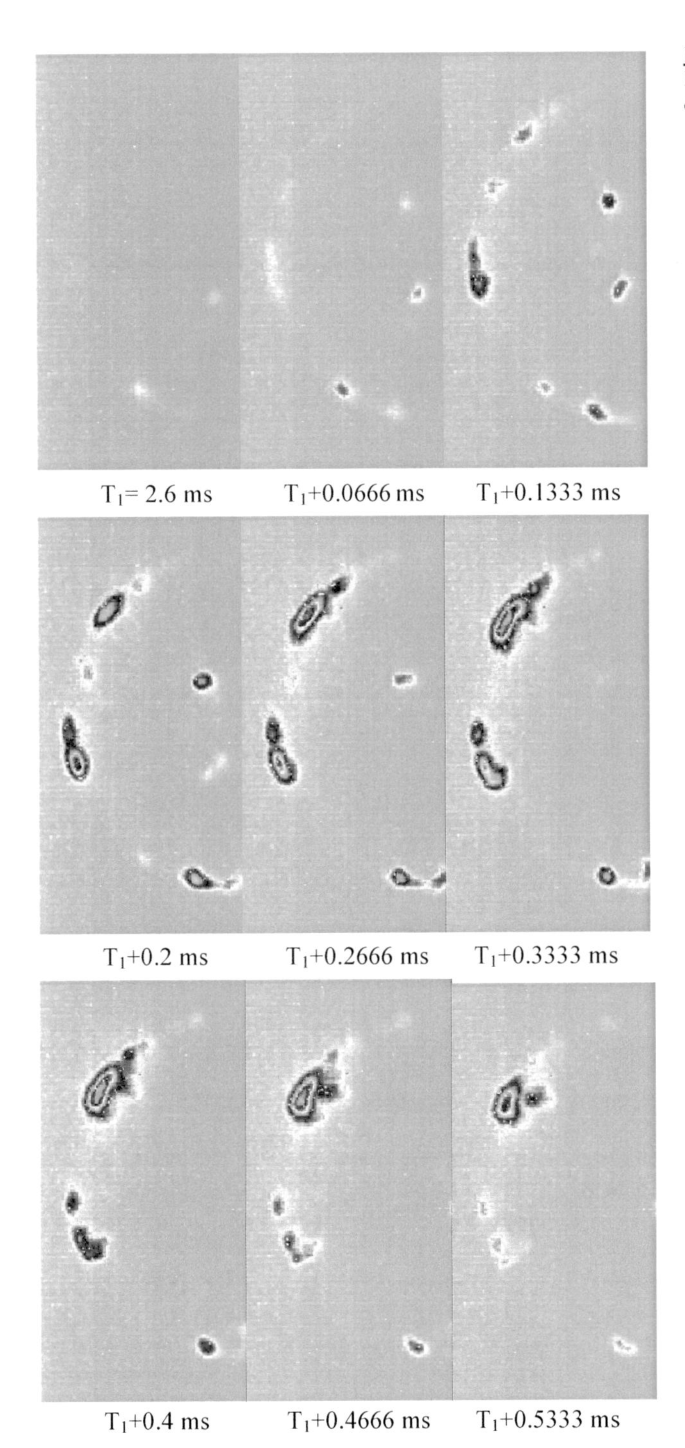

Figure 7-4:
Images of spatial soot development of SMI

Interestingly, however, is that these two locations exhibit a luminosity intensity of similar magnitude indicating that both locations experience similar combustion mode and local temperature seems to be of comparable order. The images provide no indication of diffusion combustion. As time progresses, soot develops radially and along the combustion chamber periphery, always at similar locations with respect to the fuel jet, where air-fuel ratio is at a state of favoring soot formation. As time progresses further, soot production persists only in the periphery since it undergoes oxidation relatively in short time. Most importantly, however, is that the overall soot production under these operating conditions with GTL fuel is indeed a weak process and of the characteristics of premixed combustion.

It is worth noting, however, that by comparing the start of soot appearance of this case study to that of TDC of section 6.2 (same injection timing but different injection pressure), one can reconfirm the observation made in section 6.1 that increasing injection pressure increases the portion of soot-less heat release. During this phase of heat release no soot is formed which ultimately leads to lower soot level.

7.2 Split Injection Strategy Effects

The second injection strategy investigated in this work is a *split injection* in which the total fuel mass to be injected is split into two or more injection events. This is a measure typically deployed for engine noise reduction but its consequences, in terms of combustion and soot emission characteristics, have to always be well evaluated so its merits, compared to single injection strategy, can be judged on the basis of total performance. Injection splitting considered in this work has been limited to two injection events, *pilot* and *main injection*. The pilot compromises about 25% of the total fuel mass of the main injection and its mass was kept constant throughout the investigation. Since injection timing of the pilot injection plays the key and central rule of the performance of this strategy, three cases of three different timings, referred to as *far split*, *intermediate split*, and *close split*, have been investigated. Each corresponds to a different time period separating main from pilot injection, long, intermediate, and short, correlating to 24, 12, and 6 CA° respectively. The timing of the main injection, on the other hand, has been kept unchanged and was set to the same injection timing as that of the single injection strategy in section 7.1. The far split is used as a base case for a split injection strategy whose merits are first evaluated against those of a single injection. Later, the effect of timing of pilot injection on combustion and soot characteristics is investigated by comparing the performance of the three split injection cases mentioned above.

7.2.1 Split Vs. Single Injection

Shadowgraph images were taken to first qualitatively assess the spray development with respect to the first initiation of an ignition which can be later related to a heat release pattern and, secondly, to investigate how split injection affects spray development. A set of images of the pilot injection spray up to initiation of ignition is shown in figure 7-5. The presence of flaming sites following the start of combustion of the pilot injection impeded the acquisition of any shadow images of the second (main) injection.

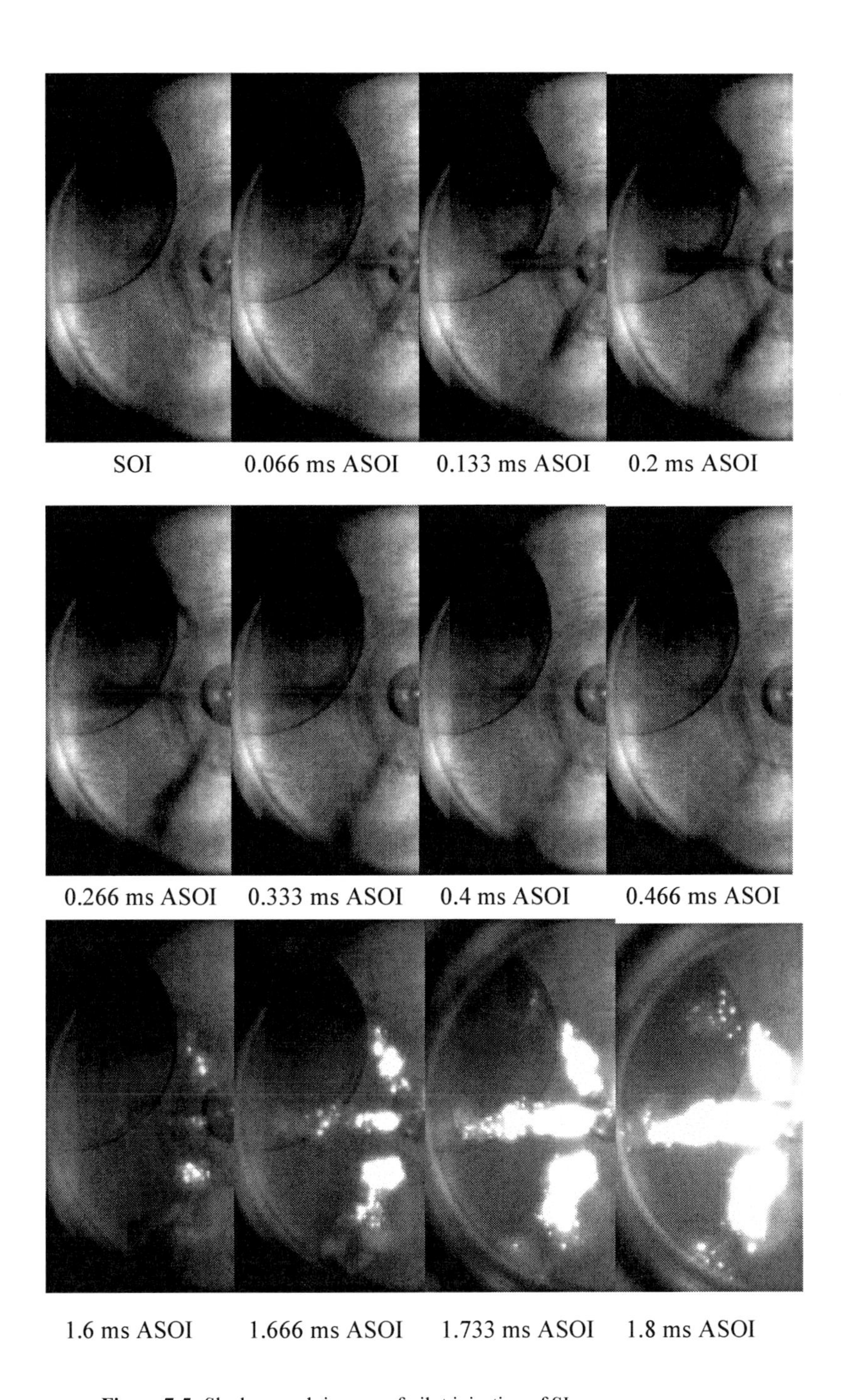

Figure 7-5: Shadowgraph images of pilot injection of SI

Visualization of spray development of the second injection, were done, therefore, by taking shadow images under suppressed combustion conditions made possible by replacing air with nitrogen which is an inert non-combustible gas. These images are shown in figure 7-6.

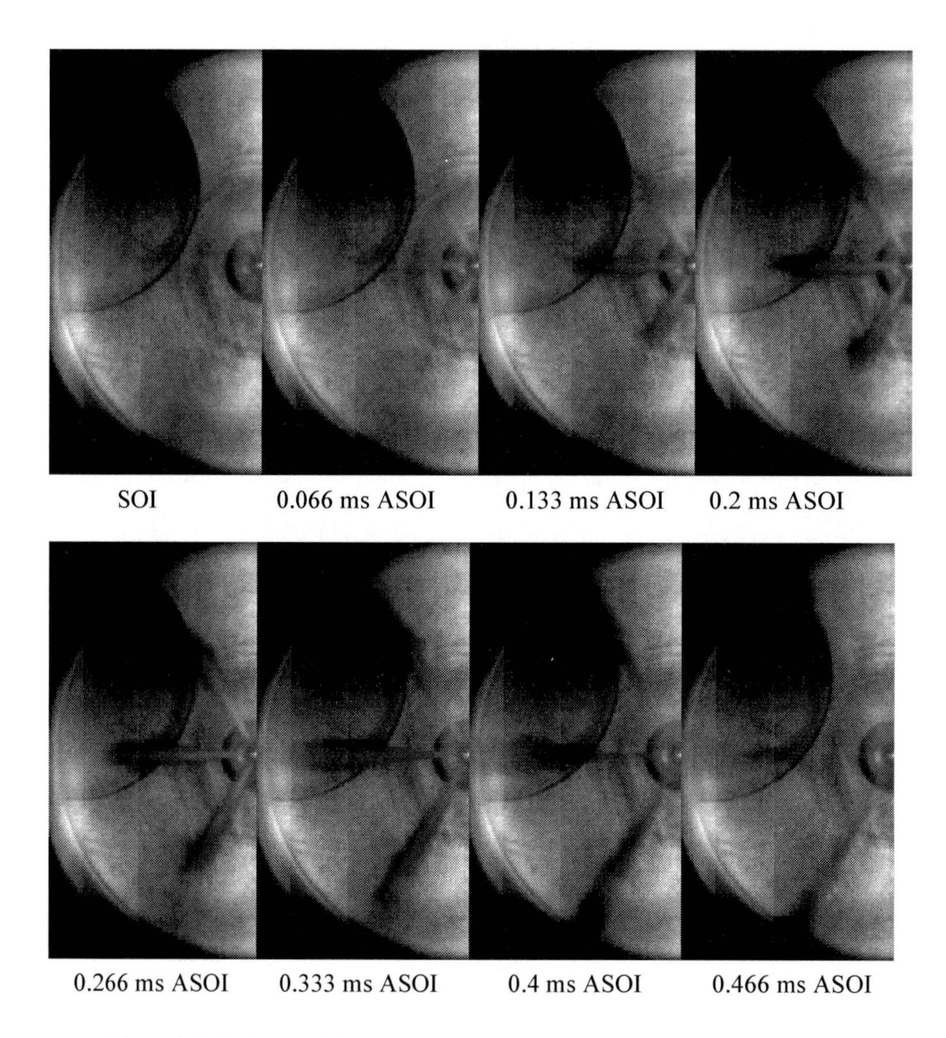

Figure 7-6: Shadowgraph images main injection of SI taken under suppressed combustion conditions

Comparing the spray images of the pilot with those of the second (main) injection of the split injection as well as with those of the single injection, one can observe the shorter penetration length the pilot spray exhibits in comparison to the second spray and that the single spray exhibits the longest penetration length. To illustrate this observation, the time-resolved spray penetration length of each one of the sprays has

been measured using software capable of tracking spray tips; the results are shown in figure 7-7. The fact that split injection exhibits shorter spray length can be attributed to the smaller momentum both first and second spray possess as a result of the splitting in injected fuel mass.
However, velocity of the sprays with split injection seems to be larger than that with single injection despite the reduced momentum they possess. The larger spray velocity which the first (pilot) spray demonstrates can be attributed to the smaller drag forces it encounters because of the lower ambient gas pressure and density. But the second (main) spray experiences somewhat a different effect which lends it a higher spray velocity. According to Lee et.al. [74], who observed a similar effect of split injection with conventional diesel spray, it seems that the first injection creates an entrained flow of ambient gas directed downstream. As a result, a relatively strong downstream flow reduces the drag acting on the subsequently injected fuel spray, thus, allowing it to penetrate faster.

The fact that split injection generates similar spray patterns for both GTL fuel and conventional diesel should not be surprising since the key fuel properties in which diesel and GTL fuel differ have no effect on physics of spray development.

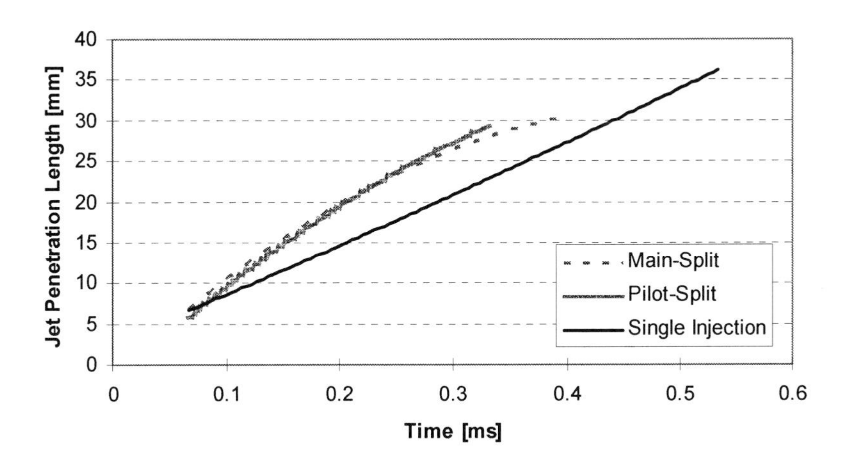

Figure 7-7: Time-resolved spray development of pilot, main-split and main-single, expressed in terms of time-resolved jet penetration length.

In order to examine the effect of split injection on combustion characteristics, the measured time-resolved cylinder pressure and the calculated ROHR have been plotted against those of single injection strategy and shown in figure 7-8.
As a result of an injection split, the rise in cylinder pressure is quite mild and occurs in two stages because heat is released in two separate stages.

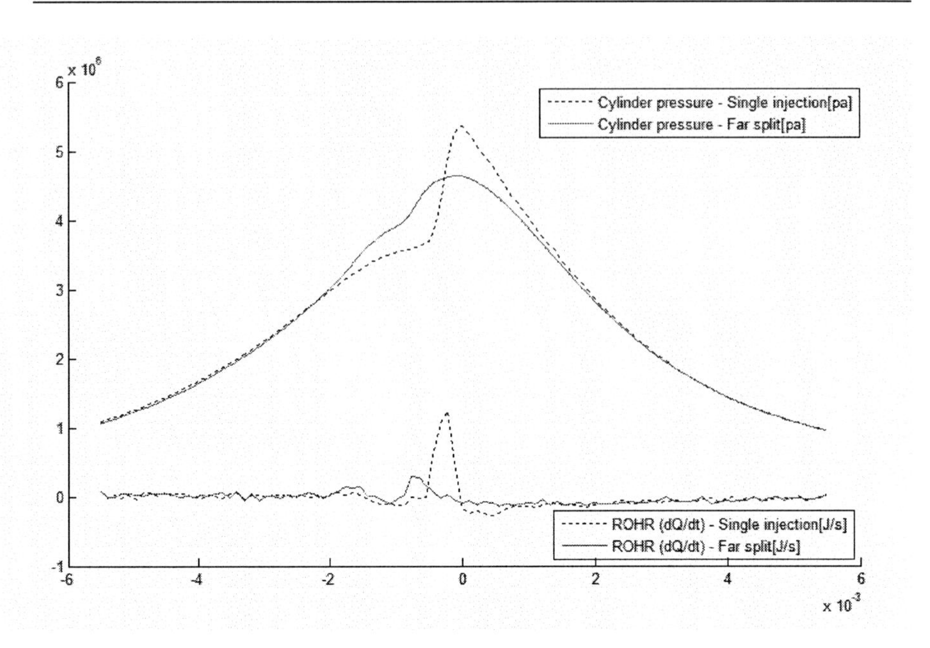

Figure 7-8: Cylinder pressure and rate of heat release curves of both single and split injection

The mild rise in cylinder pressure is also due to the shortened ignition delay of the main injection which is brought about by the heat released during first injection. The later elevates cylinder temperature and, thus, shortens the ignition delay of the second injection. As a result, the residence time available for air-fuel mixing is also shortened reducing the premixed portion of the combustion process which, in return, lessens the pressure rise rate. It is obvious from the ROHR curve that the second heat release of the split injection exhibits a mild rise which corresponds to the reduced premixed combustion portion while most of the heat is actually released in diffusion (non-premixed) combustion mode. In order to examine the mode by which the heat of the first injection is released, one needs to consider the time of flame appearance with respect to the time of heat increase. The former is obtained from the acquired shadow images shown in figure 7-5. Accordingly, the flame first appears after a sufficiently long time following completion of injection and it takes place at almost the same time of the start of heat release which indicates that the first injection burns in a premixed combustion mode with no flameless combustion portion as was the case in the single injection. Because main injection burns in a diffusion (non-premixed) mode as a result of an injection split, the heat release is, as expected, fairly declined yielding a lower combustion efficiency. Unburned Hydrocarbons and HC emissions are expected, therefore, to increase.

Furthermore, the time at which cylinder pressure reaches its peak seems to occur at almost the same time for both a single and a split injection. This is an interesting finding which indicates that split injection in the case of combustion with GTL fuel under the operating conditions of this study, requires no adjustment of combustion phasing despite the fact that ignition of the main injection with split injection starts earlier than with single injection.

In addition, the two-stage pressure rise generates a broader cylinder pressure curve with slope and peak lower than those of the single injection strategy. The combustion noise, as a result, is significantly reduced. A typical measure of combustion noise is the (dP/dt) indicator whose time evolution has been calculated and plotted for both the split, as well as the single injection cases and is shown in figure 7-9. The effect of split injection in reducing the magnitude of the (dP/dt) indicator is quite obvious. As a matter of fact, a splitting in injection, as performed in this work, has reduced the magnitude of (dP/dt) by a factor of 4.

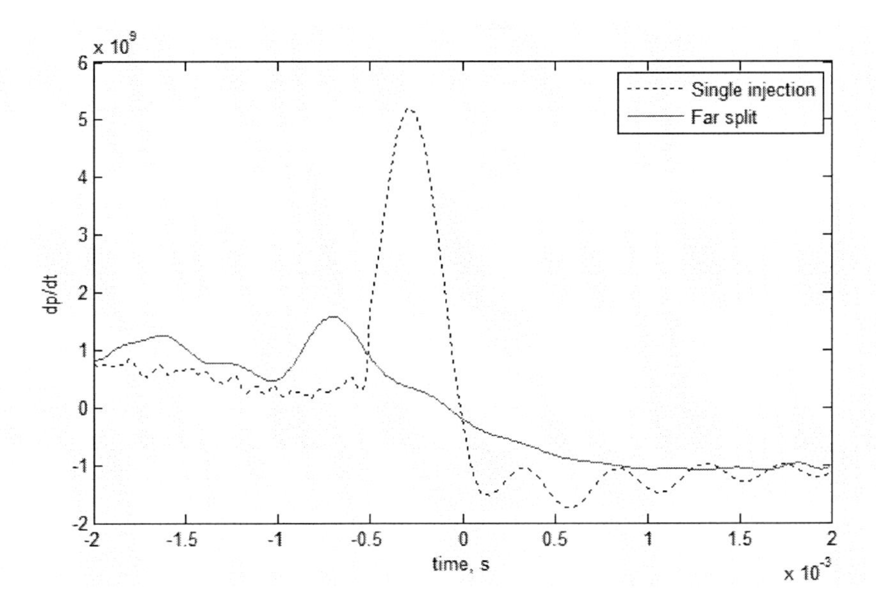

Figure 7-9: dp/dt combustion noise indicator for SMI and far split injection

The effect of split injection on soot formation has been examined through soot incandescence measurement as done in section 7.1. The temporal luminosity of soot development is shown in figure 7-10 while the spatial development of soot following soot's first appearance is shown in the set of images in figure 7-11. Accordingly, soot first appears about 1.13 ms after the first (pilot) spray fully evaporated (obtained from shadowgraph imaging) and about 0.333 ms following the onset of the premixed flame of the first injection. Unlike the case of single injection, it initiates downstream and propagates radially outward in the area where spray jets have been already entrained in air. However, the soot luminosity exhibits very weak intensity, up to a certain time in which the sudden increase following it is evident. The former indicates the stage of an extremely low soot level formed during premixed combustion of the first spray while the later marks the stage of soot formation of the second spray which initially forms within the premixed flame and later develops rapidly in the diffusion flame. This, in return, emphasizes the strong dependency of soot formation on temperature and residence time during which soot resides within the diffusion flame.

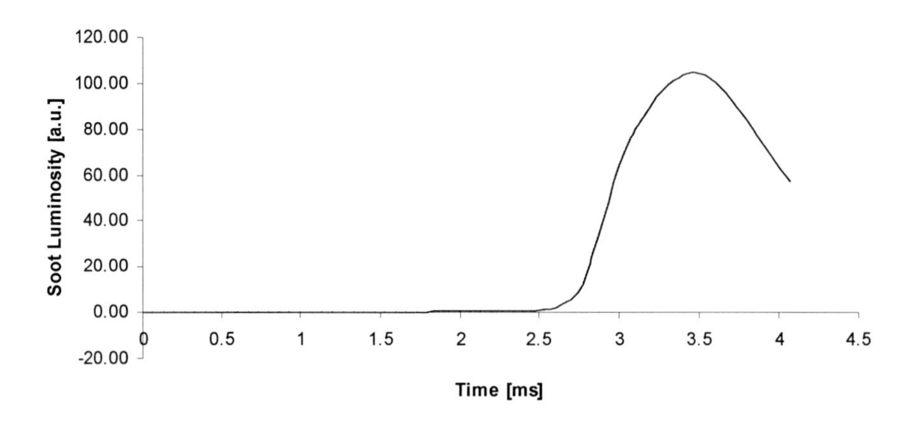

Figure 7-10: Time-resolved soot luminosity (SISI) for far split injection

Comparing the soot level with split injection to that of the single injection, one can clearly note the higher soot level produced by split injection. As a matter of fact, soot formation level with split injection is one order of magnitude larger than that of single injection. This can be attributed to the combustion mode each undergoes. Because the single injection undergoes, partially, flameless heat release and almost no diffusion combustion because the entire fuel spray fully evaporates prior to the onset of combustion, the local air-fuel ratio and local temperature are lower triggering a lesser soot formation level. Furthermore, in the case of split injection, a good segment of the second spray impinges on a burning zone, thus increasing the rate of a soot precursor production and as residence time increases the more of these precursors develop into soot. Once again GTL fuel preserves the soot formation patterns of split injection established for conventional diesel. The key, yet important difference, is that GTL fuel requires no combustion phasing as a result of deploying split injection strategy.

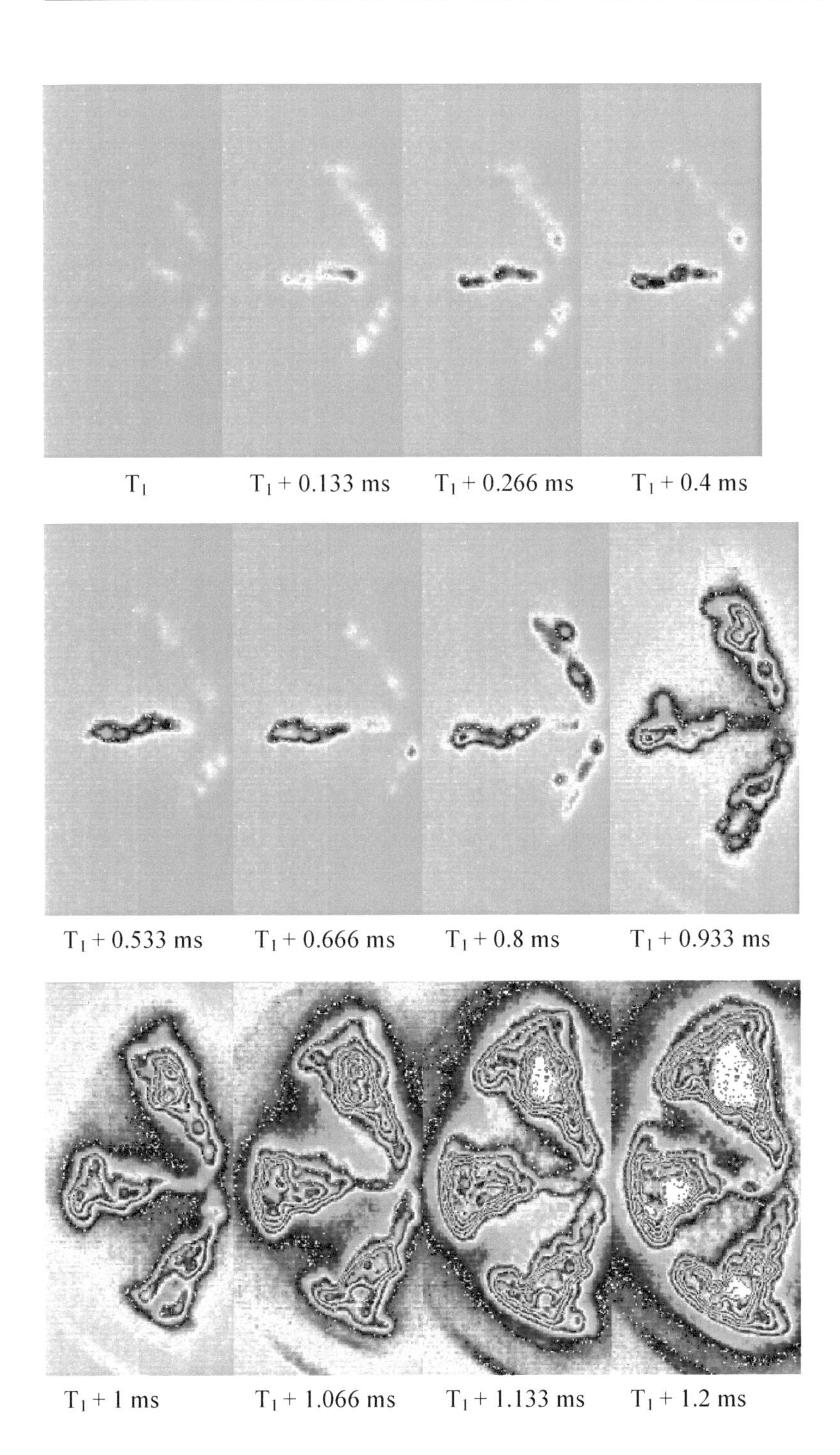

Figure 7-11:
Images of spatial soot development of far split injection

7.2.2 Effect of Pilot Injection Timing

As already mentioned, to investigate the effect of injection timing of the pilot three cases were considered. The first case, which is referred to as "Far Split", is the base case and has been already presented in the previous section. The other two cases are intermediate split and close split. In the later, the start of injection of the pilot has been brought fairly close to the main injection, with a time difference separating them equivalent to 6 CA degrees. This has been possible thanks to the fast needle opening of the piezo-injector. In the former, the start of the injection of the pilot has been set to an intermediate time, almost half the distance between far and close split case.

First, the effect of pilot injection timing on combustion and ignition characteristics is analyzed. For this purpose ROHR has been calculated as has been done up until now and results are plotted along the time-resolved cylinder pressure as shown in figure 7-12.

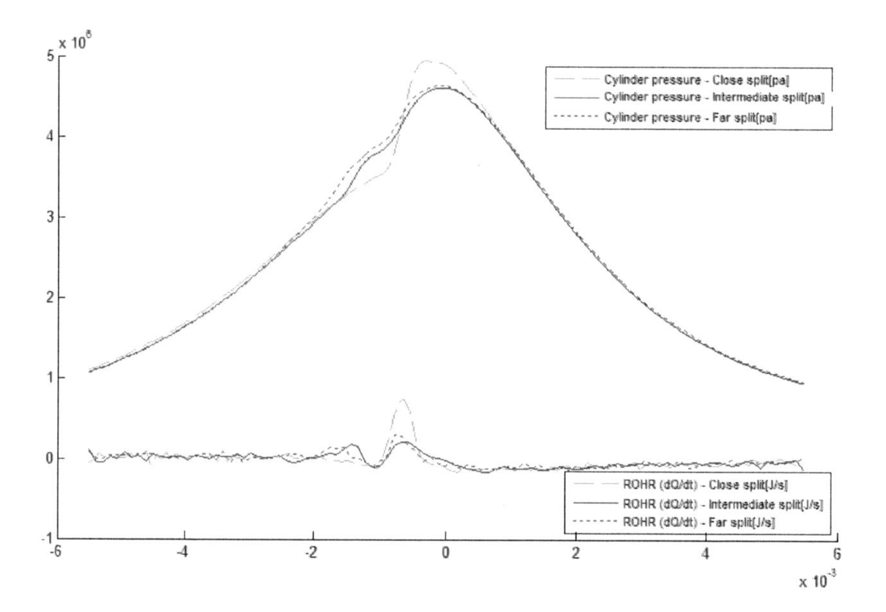

Figure 7-12: Cylinder pressure and rate of heat releases curves of far, intermmediate and close split injection

Both far and intermediate split exhibit distinct two-stage heat release resulting in two-stage pressure rise with moderate rise rate. Interestingly, however, the timing of start of pilot injection does not affect the start of the heat release of the second (main) injection. This is a very important characteristic which indicates that GTL under the operating conditions deployed in this work exhibits very stable ignition characteristics. This is mainly because by lowering CR, the rise in cylinder temperature brought about by the pilot injection is quite moderate and ignition of the second injection is controlled mainly by cetane number effects.

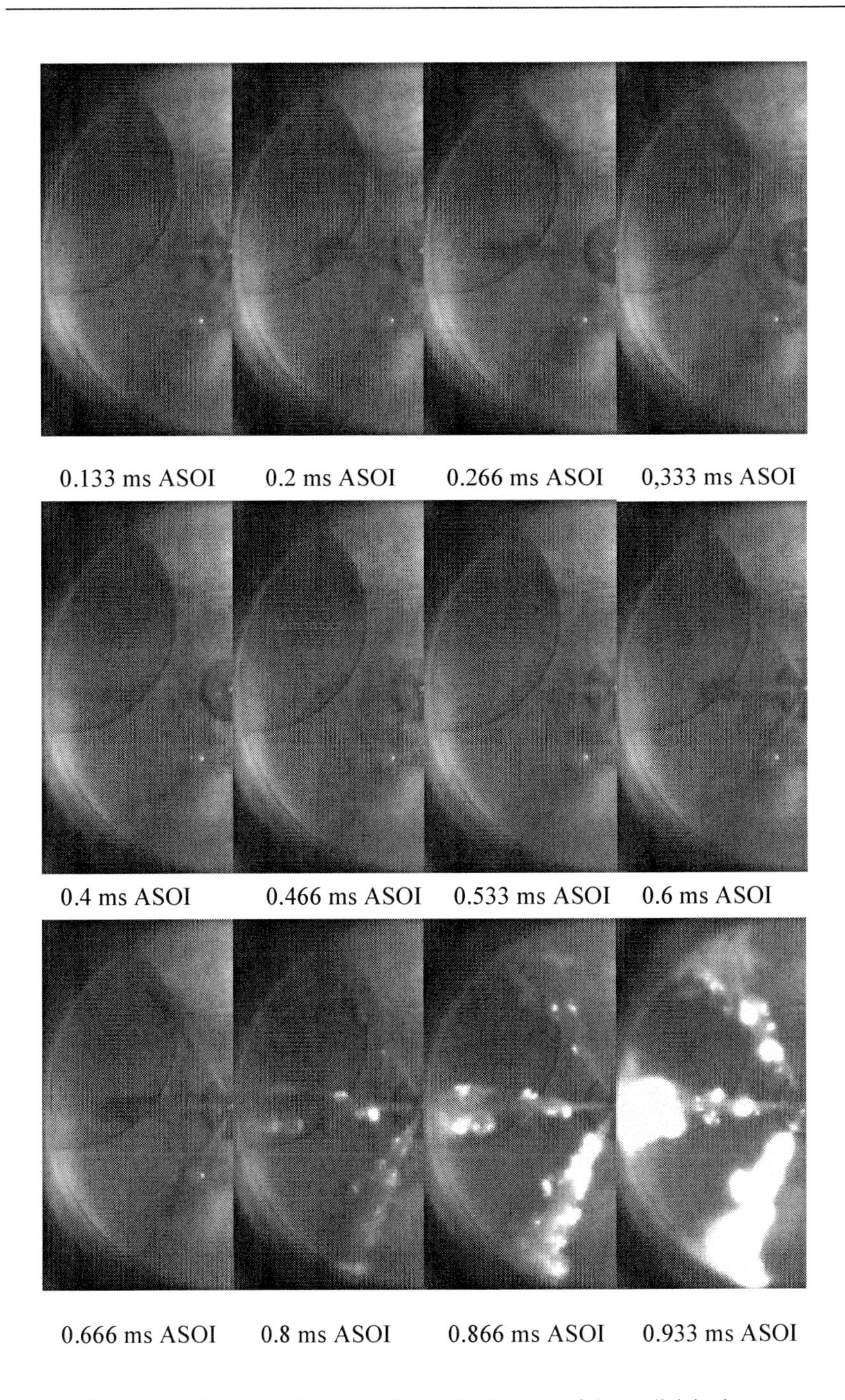

Figure 7-13: Shadowgraph images of spray development of close split injection

Such stable ignition of the second (main) injection brings the cylinder peak to occur almost at same time for all split injection cases regardless of the pilot injection timing. This, in return, eliminates the need for combustion phasing.
By further retarding the start of pilot injection, however, as was done in the case of the close split, heat release develops in a single stage and the portion of the second injection that entrains in air before the onset of ignition increases, leading to a steeper rise in both heat release and cylinder pressure.
To elaborate this, shadowgraph images were taken and a relevant set is shown in figure 7-13. It is clear that similar to the other cases, the pilot injection evaporates completely; but unlike the other cases, a sizable portion of the second injection propagates through the combustion chamber and entrains with air prior to the onset of ignition. This actually demonstrates the utilization of cooling effect in which the temperature of the cylinder's mixture drops as a result of heat loss to the evaporation of some of the fuel spray of the second (main) injection. This in return delays ignition which is translated into longer time for pre-mixing.
One can also note the shorter penetration length of the pilot injection in this case. This is because the drag on the pilot jet is larger due to the denser surrounding air. This lessens the quality of air entrainment and yields a richer premixed mixture closer to injector nozzle. Ignition, on the other hand, seems to initiate somewhat downstream of the second spray and upstream of the pilot spray reflecting the zones in which fuel has already evaporated and formed a combustible mixture.
As a result of the increased premixed combustion portion in the case of close split, the accumulated heat release is remarkably higher than those of far and intermediate split cases. Higher accumulated heat release is the result of a more complete combustion which yields a lower level of Unburned HydroCarbons [UHC].

The increased premixed combustion portion generates, on the other hand, a stronger combustion noise. The (dP/dt) indicator has been calculated for the three cases and is shown in figure 7-14.
From figure 7-14, one can clearly observe the positive rise in (dP/dt) of the main injection which occurs almost at the same time for all three split cases which, once again, indicates how the timing of a pilot injection has no effect on ignition delay of main injection. Also shown is the single stage pressure rise rate characterizing the close split case whose rate and peak are obviously higher. As a matter of fact, the peak in (dP/dt) of the close split is larger than that of the far split by a factor of 2.5 yet it is still smaller than the single injection case. Similar to the peak in heat release, the peak in (dP/dt) of the two stages of the intermediate split case is slightly higher than that of the far split but of comparable magnitude.

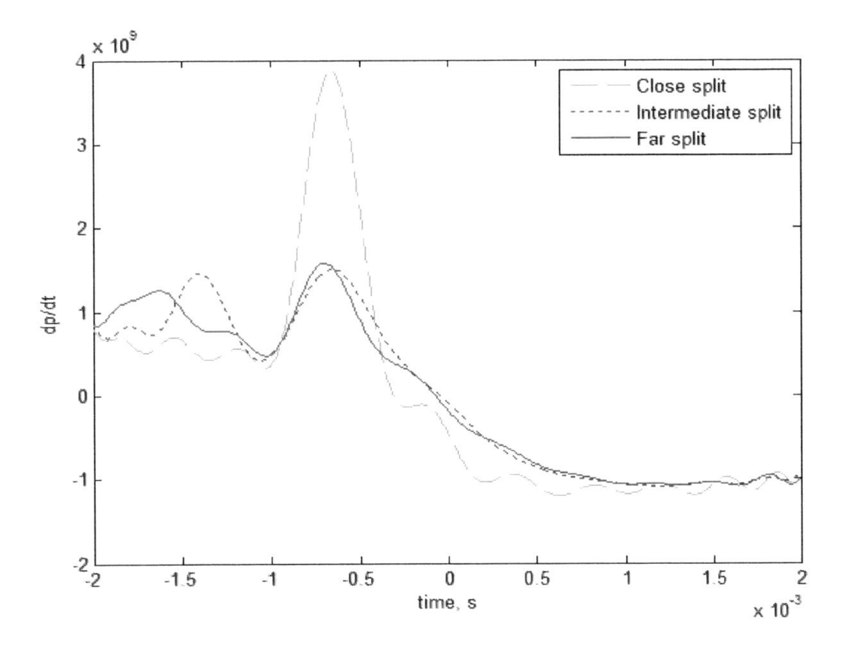

Figure 7-14: dp/dt combustion noise indicator for far, intermediate and close split injection

The effect of pilot injection timing on soot formation has been examined through the imaging of soot incandescence from which time-resolved SISI has been established as has been explained in the previous sections. Such temporal development of in-cylinder soot is shown in figure 7-15.

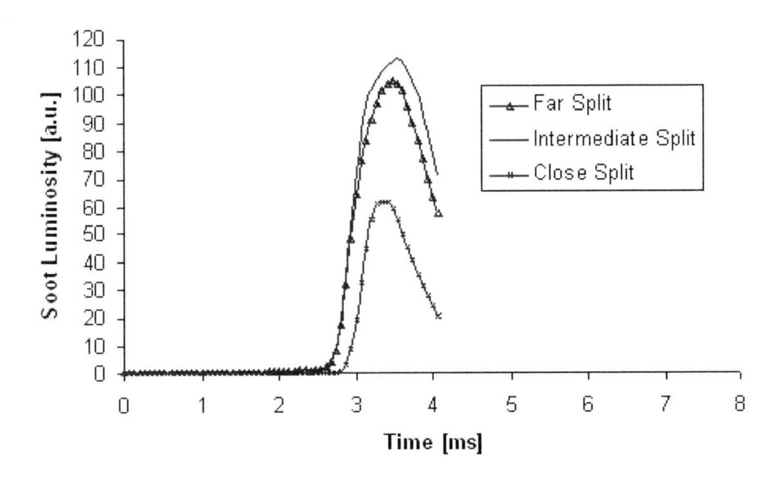

Figure 7-15: Comparison of the time-resolved SISI for far, intermediate and close split injection cases.

It is also evident here that the premixed combustion of all cases is actually soot-less. This is indicated by the very low magnitude first peak (mainly in the case of far and intermediate split) before an obvious rise in soot luminosity takes places reflecting the sooting characteristic of the diffusion (non-premixed) combustion. This shows, once again, the strong dependency of soot on temperature and local air-fuel ratio. Because premixed combustion under these operating conditions is almost soot-less, the rise in soot luminosity in the case of close split is significantly delayed and the peak in soot formation is significantly lower. This is because the premixed combustion portion, in this case, is remarkably larger than the other two cases.

To further elaborate this observation, we consider the spatial soot development of this case. A representative set of images showing spatial soot development is shown in figure 7-16. Unlike the other split cases, in the close split case a first appearance of soot takes place simultaneously downstream and upstream. The former corresponds to the premixed zone generated by the pilot injection while the later corresponds to the premixed zone of the second (main) injection which compromises mainly of the outer envelope of the second spray that has already entrained in air. These two regions exhibit very low luminosity intensity until the onset of diffusion combustion which generates larger pool of soot. Interestingly, however, is that soot formed downstream the fuel jet as a result of fuel being impinged on burning zone tends to propagate radially and further develop along the periphery of the combustion chamber.

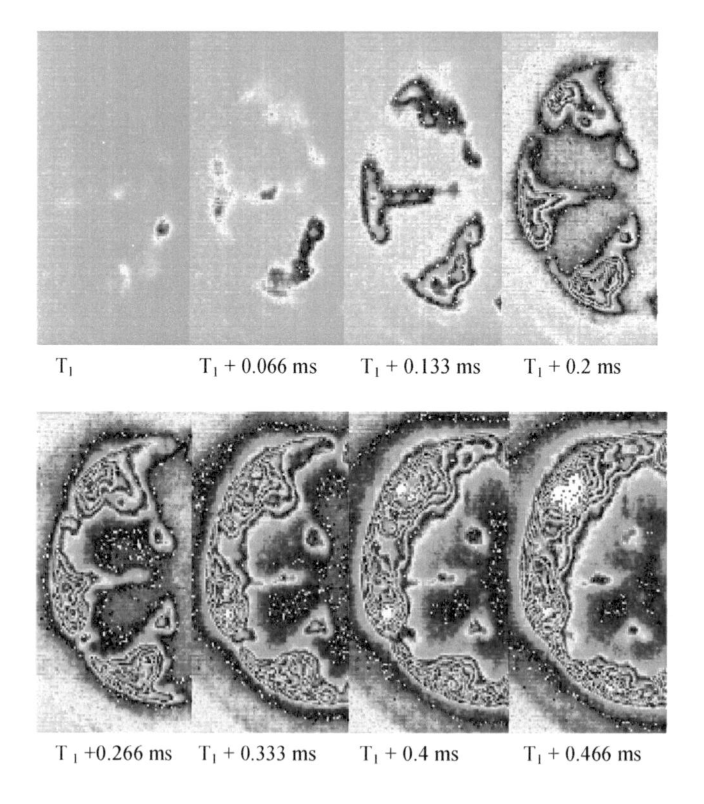

Figure 7-16: Images of spatial soot development of close split injection

Because under the operating conditions deployed in this work, pilot injection timing has no effect on ignition delay neither on the start of heat release of the second (main) injection yielding as a result no shift in the timing of cylinder pressure peak, final soot level can be correlated to the soot luminosity peak. It is used, therefore, as a measure to assess the effect of pilot injection timing or a time difference between pilot and main injection denoted as dSOI on soot formation level. Figure 7-17 shows the soot luminosity peak of the three cases.

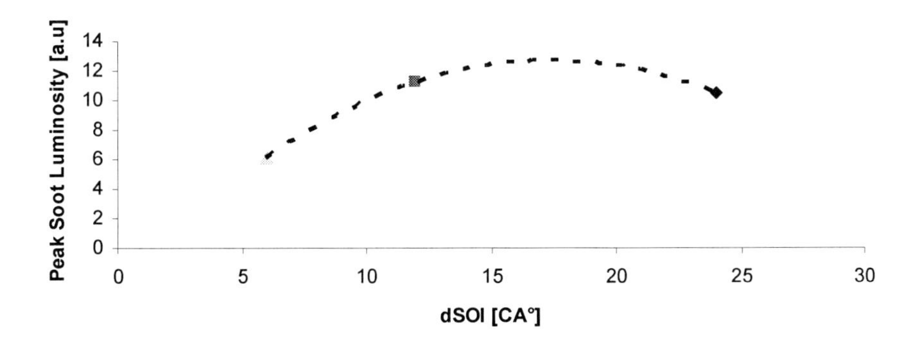

Figure 7-17: Soot luminosity peak as function of time difference between pilot and main injection dSOI

Interestingly, it shows a pattern of a quadratic relation of positive parabola, so that far and close split generate lesser soot than intermediate. Close split generates, however, the lowest soot level. In addition to the already given reasoning, one should consider a spatial characteristic as it pertains to the local air-fuel ratio in order to better understand such behavior. Since, as has been shown up to now, the Lift-Off Length [LOL] is indeed a useful correlation to local air-fuel ratio (local mixing state) and soot production tendency is strongly depending on it, an assessment of the Lift-Off Length from the acquired soot incandescence images has been drawn but in this case it is referred to as LOL of sooting zones rather than of flame zones.
The measured LOL of soot has been plotted and shown in figure 7-18. It, together with figure 7-17, shows that the peak in soot incandescence is inversely proportional to soot LOL. The longer the soot LOL is, the farther soot formation sites from injector nozzle are and, thus, soot is formed at leaner mixture conditions generating, as a result, an overall lower soot level. This means that the soot for the case of close split is formed at the leanest mixture conditions, while for the intermediate split case it is produced at the richest mixture conditions. In other words, varying the timing of pilot injection or the time difference between pilot and main injection directly affects the in-cylinder mixture formation. This brings about a different air-fuel premixing quality which, in return, yields a different soot LOL ultimately resulting in different overall soot levels.

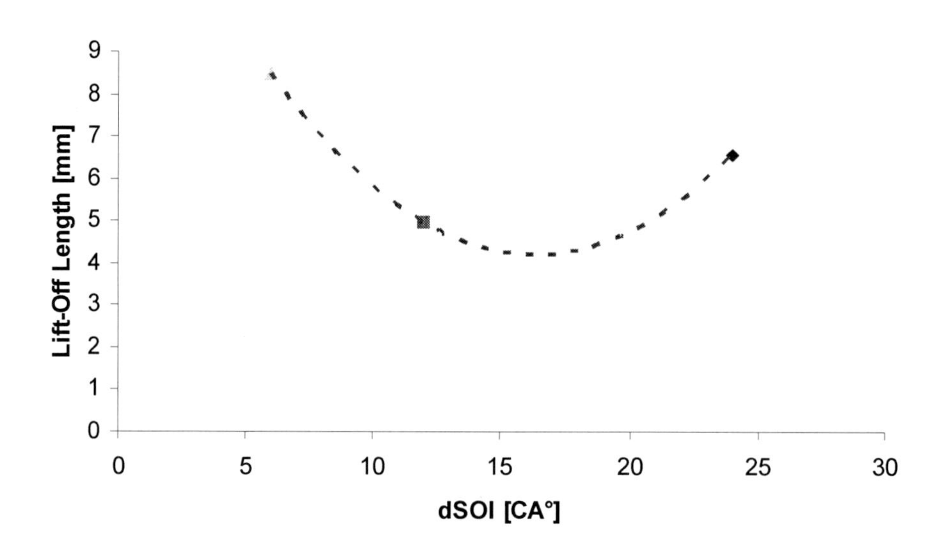

Figure 7-18: LOL as function of time difference between pilot and main injection dSOI

The lowest soot level brought about by close split injection makes split injection of the pattern of close injection events certainly favorable. Favoring close split injection over the rest of split injection cases is also due to the benefits it brings about in terms of reduction in combustion noise, yet maintaining heat release rate, combustion efficiency relatively close to those of single injection. Comparing the performance of close split injection to that of single injection by considering the already presented results and analysis one can claim that because close split injection yields reduction in combustion noise in a rate higher than 25 %, peak cylinder pressure and peak heat release rate lower than that of single injection by less than 10 % and 30 % respectively, yet more than 50 % reduction in peak soot concentration as compared to the other split injection cases, one can claim that in the event of deployment of split injection, close split injection is indeed the favorable arrangement of injection events. This finding is, however, somewhat different from the observations made by Vanegas et.al. [72] who showed that by applying split injection strategy in the case of conventional diesel fuel, a far split injection has produced the lowest smoke emissions and generated longer ignition delay of the main injection while intermediate split yielded highest heat release. The fact that GTL fuel exhibits stable ignition and as a result the time difference between pilot and main has no effect on ignition delay of the main injection could very well be the reason behind such difference.

7.3 Multiple Injection Strategy Effects

The multiple injection strategy investigated in this work is comprised of three injection events which is a combination of the first two events of the split injection strategy and a third injection typically referred to as "*Post Injection*".
Such an arrangement of injection events is better listed under multiple injection strategy because in the split injection strategy, split of injection, which can be of multiple injection events, is deployed to alter the combustion process in a manner of splitting to large extent the heat release as already explained. The additional post injection does not alter the combustion process, nor does it contribute to the split of a heat release. It is merely a measure to influence the post combustion processes and chemistry aiming mainly at altering the soot formation mechanism with the intention to reduce the final soot level. Split injection is used in the context of this work, therefore, as the strategy which alters the combustion process with respect to single injection regardless of the number of injection events. On the other hand, multiple injection is used in the context of this work as a strategy which encompasses post combustion injection events.

Since the timing of the post injection is the key parameter that affects the final soot level, three cases of three different timings, referred to as, *close*, *intermediate*, and *far post-injection* corresponding to a distance from the main injection equivalent to 12,15, and 18 CA° respectively were investigated. The first two events of the split injection were kept unchanged. The later were those of the base split injection case, hence of the far-split case. The fuel amount injected in the post injection was kept unchanged and equals to about 10% of the main injection.

Investigating the effect of post injection timing on soot formation is being conducted in the light of the already established theories as described in section 2.3.3. It is worth noting, however, that the author dismisses the theories given by [82, 83, 84] since assessment of the effect and merits of post injection should be done without altering the rest of the injection events; basically, without changing the amount of fuel added to the system which determines the heat released during main combustion. It is the convention of the author that analysis should be done at constant work (or torque) output as that of the split injection. If to examine the effect of post injection at constant mass as they suggest, hence the amount of fuel injected in the post injection is extracted from the main injection, then it is trivial that the main injection would produce lesser soot and the addition of post injection in this case neither further reduces soot nor does it improve combustion and, thus, one needs not to deploy it in the first place. Deployment of post injection is sinful, therefore, only if it can reduce soot level of that case without post injection.

Furthermore, analysis is based on the assessment of soot formation of the post-injection itself. The evaluation criteria is whether post-injection of the three cases produces soot and at what magnitude.

Here as well, the soot volume fraction has been measured by high speed imaging of soot incandescence from which the time-resolved SISI has been extracted.
Such temporal soot evolution of the three investigated cases is shown in figure 7-19.
It is clear that all post-injection cases temporarily produce an additional amount of soot while each interacts differently.

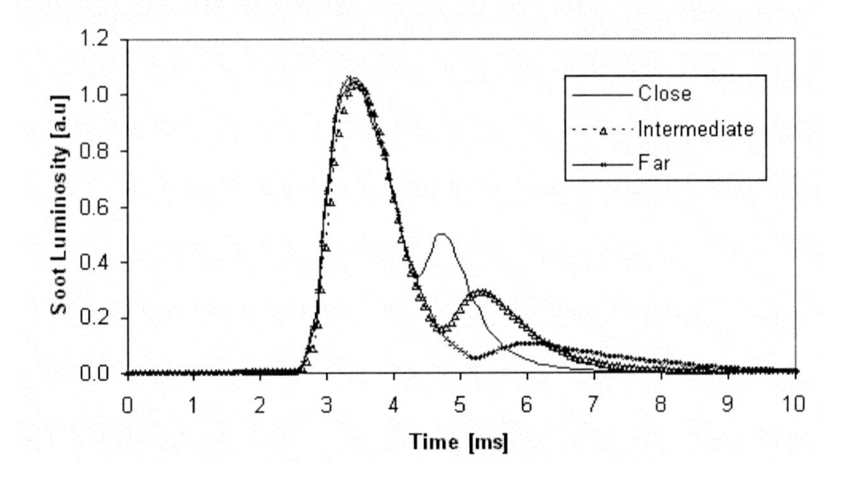

Figure 7-19: Time-resolved SISI of far, intermediate and close post-injection

The close post-injection, although temporarily, produces a significant amount of soot, exhibits a faster rate of soot destruction or burn out. The narrow curve of the second soot luminosity peak, characterized by a distinct rapid falling slope, is a strong indication of rapid soot destruction brought about by strong activity of soot oxidation. Post-injection taking place during combustion or very close to the end of combustion enhances the burn of soot particles [71]. The extra energy brought about by close post-injection, whether in terms of thermal or mixing energy, accelerates mixing and, combined with higher temperature increases, rate of soot oxidation. Figure 7-20 shows how the post-injection jets interact with the surrounding sooting zones. Comparing it with far post-injection, figure 7-21, one notes that in the far-post injection case, two distinct split sooting zones are present. These two zones correspond to the split flame phenomenon where two distinct non-interacting flames/sooting zones are generated, one from the combustion event associated with the main injection and a second zone generated by the post-injection. Because of the absence of interaction between the post-injection pulse and the combustion zones of the main injection, very little mixing takes places, if at all, resulting in very weak soot destruction rate.

The far post-injection, on the other hand, produces actually little soot because of the lower temperature at the time of injection. These, in return, confirm first that soot formation requires certain temperature below which soot either does not form at all or forms at very low level. Because cylinder temperature in the case of close post-injection was relatively high, soot formed in sizeable amount while in the case of far post-injection where temperature is relatively low soot formation level was fairly low. Second, that for post-injection to have an influence on soot destruction, both higher temperature and mixing energy should be added to the system in order for both to accelerate soot oxidation. In other words, it is not sufficient that the additional post-injection does not produce significant amount of soot as the case of far post-injection. Rather, it is essential that post injection promotes soot oxidation generated by the main combustion process. This necessitates apparently a combination of higher temperature brought about by addition of heat provided to the system through an injection of extra amount of fuel at combustible condition as well as extra mixing energy brought about by jet pulse capable of transporting reactants to the sooting sites of the main combustion process which in return increase the rate of the carbon oxidation reactions. These seem to be realized through an early post injection timing.

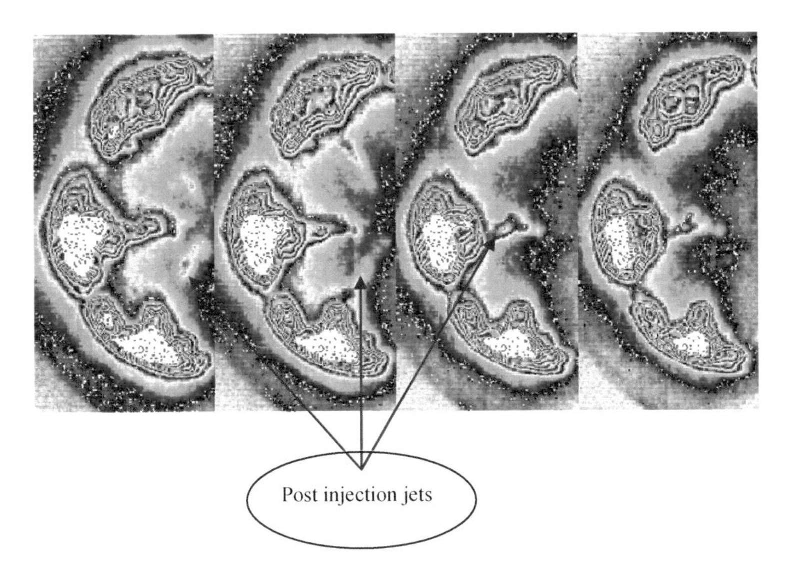

Figure 7-20: Images of soot incandescence showing interaction of post-injection spray with surrounding soot from main combustion (close post-injection case)

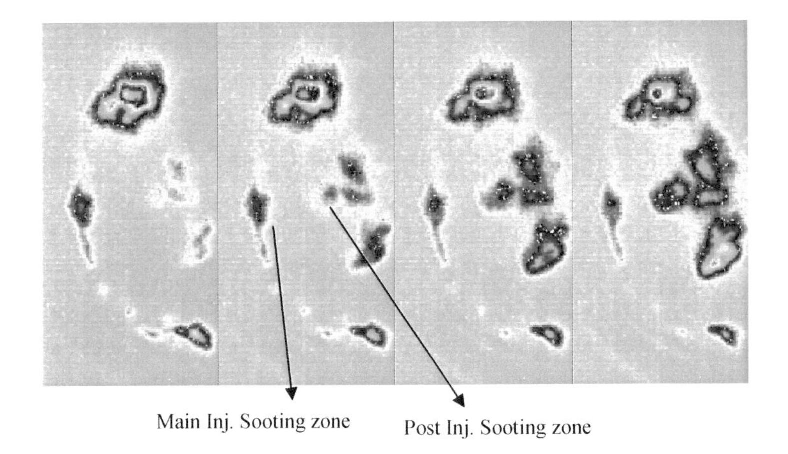

Figure 7-21: Images of soot incandescence showing separated non-interacting soot zones (case of far post-injection case)

To further elaborate this, the temporal evolution of OH chemiluminescence [Chem.], which is used as indicator of burn and oxidation activity, has been measured and is shown in figure 7-22. Although a comparable magnitude of increase in OH Chem. for close and intermediate post-injection has been measured, the OH production level of the close post-injection is significantly higher than that of the intermediate as deduced from figure 7-22. This indicates a stronger oxidation activity and explains why the rate of soot oxidation in the case of close post-injection is fairly higher than that of intermediate case. Furthermore, although the pool of OH in the case of far post-injection is sizable, neither the rate of soot production nor the rate of soot destruction is remarkably high, emphasizing once again the strong dependency of soot formation and soot oxidation on temperature as well as on mixing and interaction between the post injection jet and the sooting or burning zones of the main combustion (corresponding to the combustion of the main injection). So long temperature is low, both soot formation and soot oxidation rates are consequently low.

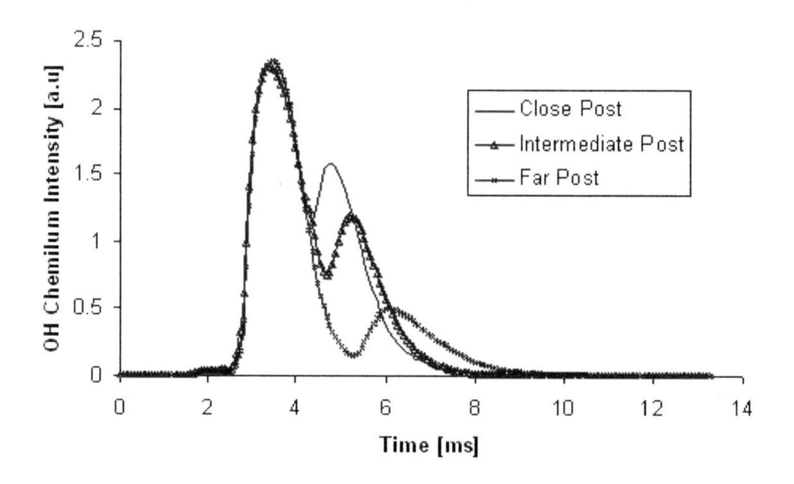

Figure 7-22: Time-resolved OH chemiluminescence of close, intermediate and far post-injection cases

Comparing the three cases of multiple injection with the base case of split injection, it seems that under the operating conditions deployed in this work and the utilization of GTL fuel, post-injection would probably produce an additional amount of soot and overall soot level would probably be higher than that without post-injection. This could be attributed to the lower sooting tendency of GTL fuel and under the operating conditions deployed in this work, mainly low CR, fuel properties take over all other effects.

This assessment cannot serve, however, as a final conclusion in regards to the post-injection effect on final soot level, especially because the validity of the deployed soot volume fraction measurement technique is limited to the early stage of the expansion stroke.

The full soot formation or oxidation pattern is not captured as a result. Soot or smoke measurement at the exhaust should, therefore, be performed in order to verify or confirm this finding.

8 Summary

This work capitalized on the investigation of the effect of physical parameters namely compression temperature and pressure as well as injection parameters namely injection pressure and timing in addition to the effect of injection strategy namely single, split and multiple injection on the combustion and in-cylinder soot formation performance of Gas-To-Liquid fuel. A fundamental objective of conducting such investigations was to determine what combustion mode is realized through the variation of such parameters. Several diagnostics techniques were deployed for this purpose. Cycle-resolved cylinder pressure and volume for combustion and heat release analysis, shadowgraph imaging for the visualization of in-cylinder processes namely spray development and ignition zones, imaging and non-imaging cycle-resolved Soot Incandescence for the determination of temporal and spatial soot development as well as OH-Chemiluminescence for the time-resolved qualitative measurement of OH concentration.

It has been shown that lowering compression temperature from 847 to 813 K through reducing Compression Ratio [CR] from 16 to 14 allowed prolongation of ignition delay by about 500 µs. Through this the injected amount of fuel has been fully evaporated before the onset of combustion. Thus the combination of GTL fuel and low CR guaranteed the realization of a combustion mode which was titled in this work as Low Compression Ratio Pre-mixed [LCRP] Combustion. Furthermore, when introducing fuel into the combustion chamber in a single injection, heat has been released in two distinct modes, flameless as well as low sooting pre-mixed flame.

In the case of GTL fuel like the case of conventional diesel fuel, Lift-Off-Length has the strongest effect on soot formation. The longer LOL is the leaner the mixture is and the lesser soot forms. Effects of key parameters such as cylinder and injection pressures on soot formation are reflected in the effect of these parameters on LOL. Effect of cylinder pressure (air density), on soot formation exhibits negative quadratic relation while injection pressure effect exhibits a linear relation. So increase in injection pressure results in increase in LOL and ultimately decrease in soot formation while increase in cylinder pressure (air density) results in decrease in LOL and ultimately increase in soot formation. Increase of air density on the other hand increases the oxygen content in the mixture which should contribute to the acceleration of soot oxidation and burn out. Increasing cylinder charge pressure to 1.5 [bar] did not produce, however, soot level lower than the case of cylinder charge pressure of 1.05 despite a higher soot oxidation rate.
Furthermore, increasing injection pressure increased the percentage of soot-less heat release which in return reduced the overall soot formation level. Because of the relatively lower compressibility of GTL fuel as a result of its lower density, injection pressure had to be elevated to high level, as high as 1600 [bar], in order to for a reduction in soot formation rate to be tangible.

Injection timing had a quite interesting effect on soot formation. Retarding injection timing brought cylinder pressure peak to occur slightly after top dead center and thus portion of the heat was released during expansion stroke. This in return accelerated soot oxidation due to higher thermal and mixing energy and brought soot formation level to its lowest.

Late injection in the case of combination of GTL fuel and low CR seems to effectively allow the utilization of cylinder charge cooling effect which in return lowers mixture temperature resulting in lower soot formation level. Late injection like high injection pressure increased as well the percentage of soot-less (also flame-less) heat release. As a result the combination of GTL fuel and low CR together with late injection and sufficiently high injection pressure (higher than 1300 bars) allowed the realization of the so called Highly Premixed Late Injection [HPLI] or HCCI-like combustion mode.

Injection strategy and injection rate shaping have a remarkable effect on combustion and soot formation of a GTL fuel. Like conventional diesel fuel split injection alters the spray development and mixture formation. The split in jet's momentum yielded shorter penetration length of both injection events, yet lent them higher velocity. It has altered the combustion mode from fully premixed combustion as in the case of single injection to two different modes, a premixed combustion of the pilot injection followed by non-premixed (diffusion) combustion of the second injection event. This on one hand has lowered combustion noise but has on the other shortened ignition delay and yielded soot formation level one order of magnitude higher than single injection. Split injection has also lowered the accumulative heat release and subsequently lowered combustion efficiency. Unburned Hydrocarbons and HC emissions, should be expected, as a result, to be fairly higher than those of single injection strategy.
The time difference between both injection events, hence between pilot (first injection event) and main injection (second injection event) has been shown to be the key parameter in determining the performance of split injection. The fundamental finding is, however, that GTL fuel under the conditions of low CR exhibits very stable ignition and as a result the time difference between both injection events does not affect the ignition delay of the main injection. This brought the peak in cylinder pressure to occur almost at same time requiring no combustion phasing.

Another remarkable finding that distinguished GTL fuel from conventional diesel fuel is that close split or short time difference between injection events yielded the lowest soot formation level despite higher peak in heat release rate. The effect of time difference between split and main injection on soot formation level has been found to have a positive quadratic relation where intermediate split injection produced the highest soot level while far and close split produced lower levels. This has been attributed to the effect of LOL on soot formation. LOL exhibited a quadratic relation as well, yet a negative one so intermediate split injection yielded the shortest LOL while far and close split yielded the longest LOL. Close split has also altered the combustion mode of the main injection from fully non-premixed to a mix of premixed and non-premixed combustion. As a result the heat release rate was higher than the rest of the split cases and was shown to be of comparable order of a single injection. Combustion noise, although lower than single injection, was the highest of all split cases. Nevertheless, such split injection arrangement, namely, close split characterized by short time difference between injection events seems to be of a favorable strategy. It effectively utilizes the cooling effect of the charge mixture brought about by the heat lost from the mixture to the evaporation of the subsequent fuel jet which in return enhances fuel air pre-mixing. As a result the trade-off between combustion efficiency or output work, combustion noise and soot formation is somewhat relaxed.

It is the convention of the author, therefore, that deploying advance injection strategies such as multiple close splits and boot injection as well as injection with variable injection pressure combined with operation on GTL fuel could very well resolve completely such trade-off and offer optimal overall performance.

The effect of post injection timing as part of a multiple injection strategy (Polit-Main-Post) on soot formation has also been investigated. It has been found that post injection regardless of its timing produced certain amount of soot. The final soot level was, however, the balance between soot production and soot oxidation. Both required high temperature but the later required a sufficient mixing energy so the post injection interacts with the combustion or soot formation sites of the main injection. This has been realized through a close post injection where post injection is injected early rather than late, producing as a result higher soot oxidation rates and ultimately lower overall soot level. This has been confirmed through a measurement of OH chemiluminescence which has shown high OH concentration compared to lower concentrations for the other cases, intermediate and far post-injection.

In general, in the case of GTL fuel under the operating conditions deployed in this work, mainly low CR, post injection does not seem to effectively contribute to lowering soot formation level in comparison to split injection without post injection. This finding needs to be confirmed, however, through smoke level measurement at the exhaust of a real engine.

8 Zusammenfassung

Dieser Arbeit beschäftigt sich mit Untersuchung der Wirkung von physikalischen Parametern nämlich Kompressionstemperatur und Druck sowie Einspritzparameter nämlich Einspritzdruck und Einspritz-Timing zuzüglich der Wirkung der Einspritzstrategie nämlich Einzeln-, Split- und Mehrfacheinspritzung auf die Verbrennung und die Innenmotorische Rußbildung des Gas-To-Liquid [GTL] Kraftstoffs. Ein grundlegendes Ziel solcher Untersuchungen war zu bestimmen, welches Verbrennungsmodus durch die Veränderung von solchen Parametern realisiert ist. Mehrere Diagnoseverfahren wurden für diesen Zweck eingesetzt. Zyklusaufgelöster Zylinderdruck und Volumen für Verbrennungs- und Verbrennungsverlaufsanalyse, Schattenrißverfahren für die Vorstellung von Innenmotorischen Prozesse nämlich Innenmotorischer Sprayverlauf und Zündungszonen, abbildende und nichtabbildende Zyklusaufgelöste Ruß-Eigenleuchten (RußInkandeszenz) für die Bestimmung des zeitlichen und räumlichen Entwicklung des Rußes sowie OH-Chemilumineszenz für die Zeitaufgelöste qualitative Messung von OH-Konzentration.

Es ist gezeigt worden, dass Absenkung der Kompressionstemperatur von 847 zu 813 K durch Reduzierung des Verdichtungsverhältnisses [CR] von 16 zu 14 eine Verlängerung des Zündverzugs um ungefähr 500 µS erlaubte. Dadurch wurde die eingespritzte Menge des Kraftstoffs vor dem Anfang der Verbrennung völlig verdunstet. Folglich, die Kombination des GTL-Kraftstoffs und niedriger CR die Realisierung eines Verbrennungsmodus garantiert, der in diesem Werk als Low CR Vorgemischte Verbrennung [LCRP] genannt wurde. Ferner beim Einführen des Kraftstoffs in die Verbrennungskammer in einer einzelnen Einspritzung wird Wärme in zwei abgesonderten Modi abgegeben, ohne offener Flamme sowie rußarme Vorgemischte Flamme.

Im Fall des GTL Kraftstoffs wie im Fall des konventionellen Dieselkraftstoffs, hat Lift-Off-Length die stärkste Wirkung auf Rußbildung. Je längere LOL desto magerere Mischung und weniger Rußbildung. Einfluss von Schlüsselparametern wie zum Beispiel Zylinder- und Einspritzdrücke auf Rußbildung sind im Einfluss dieser Parameter auf LOL reflektiert. Einfluss des Zylinderdrucks (Luftdichtheit) weist negatives quadratisches Verhältnis während Einfluss des Einspritzdrucks ein lineares Verhältnis auf Rußbildung auf. So resultiert Erhöhung des Einspritzdrucks in Verlängerung der LOL und letzten Endes Abnahme in Rußbildung während Erhöhung des Zylinderdrucks (Luftdichtheit) in Verkürzung der LOL und letzten Endes Zunahme in Rußbildung. Erhöhung der Luftdichtheit vermehrt andererseits den Sauerstoffinhalt in der Mischung, die zur Beschleunigung von Rußoxydation und Rußverbrennung beitragen sollte. Erhöhung des Ladedrucks auf 1,5 [bar] hat jedoch nicht in geringere Rußbildung resultiert als im Fall von Ladedruck von 1,05 trotz einer höheren Rußsoxydationsrate.
Ferner die Erhöhung des Einspritzdrucks hat das Prozent des rußarmen Verbrennungsverlaufs erhöht, die dafür das gesamte Level der Rußbildung verringert hat. Wegen der verhältnismäßig niedrigeren Komprimierbarkeit des GTL-Kraftstoffs infolge seiner niedrigeren Dichte musste Einspritzdruck höher eingestellt werden, (1600 [bar]), damit Reduzierung der Rußbildungsrate fassbar sein wird.

Einspritz-Timing hat ziemlich interessanten Einfluss auf Rußbildung gehabt. Verzögerung des Einspritzzeitpunktes hat Zylinderdruckspitze etwas nach dem oberen Totpunkt gebracht und folglich wurde Anteil der Wärme während des Arbeitshubes freigegeben. Dafür wurde Rußoxydation auf Grund höher Wärme- und Mischungsenergie beschleunigt und Rußbildungsrate zum niedrigsten Wert gebracht. Späte Einspritzung im Fall der Kombination des GTL Kraftstoffs und niedriger CR scheint effektiv die Verwendung von Ladeluftkühlung zu erlauben, die dafür Mischungstemperatur senkt, die in niedrigere Rußbildungsrate resultiert. Späte Einspritzung wie hoher Einspritzdruck haben ebenso das Prozent von rußarmen (auch flammenlosen) Verbrennungsverlauf erhöht. Demzufolge, die Kombination des GTL Kraftstoffs und niedriger CR zusammen mit später Einspritzung und ausreichend hohem Einspritzdruck (höheren als 1300 bar) hat die Realisierung der sogenannten Highly Premixed Late Injection [HPLI] oder HCCI-ähnliche Verbrennung erlaubt.

Einspritzstrategie und Einspritzverlauf haben einen bemerkenswerten Einfluss auf Verbrennung und Rußbildung eines GTL-Kraftstoffs. Wie konventioneller Dieselkraftstoff verändert Split-Einspritzung den Sprayverlauf und Gemischbildung . Das Moment des gesplitteten Strahles hat kürzere Eindringtiefe von beiden Einspritzereignissen ergeben, trotzdem hat ihnen höhere Geschwindigkeit geliehen. Es hat den Verbrennungsmodus von völlig vorgemischter Verbrennung als im Fall der einzelner Einspritzung zu zwei verschiedenen Modi verändert: eine vorgemischte Verbrennung der Voreinspritzung, die von nicht-vorgemischten Verbrennung (Diffusionsverbrennung) vom zweiten Einspritzereignis gefolgt wird. Dies hat zum einen Verbrennungsgeräusch abgesenkt aber zum andern Zündverzug verkürzt und ergab Rußbildung eben eine Größenordnung höher als einzelne Einspritzung. Split-Einspritzung hat auch den kumulativen Verbrennungsverlauf und der anschließend niedrigere Verbrennungswirkungsgrad abgesenkt. Unverbrannte Kohlenwasserstoffe und HC Emission sollte demzufolge ziemlich höher erwartet werden, als bei der Einzeleinspritzungsstrategie.
Die Zeitdifferenz zwischen beiden Einspritzereignissen, also zwischen Voreinspritzung (erstes Einspritzereignis) und Haupteinspritzung (zweites Einspritzereignis) hat sich als Schlüsselparameter beim Bestimmen der Leistung der Split-Einspritzung erwiesen. Das grundlegende Erkenntnis ist aber, dass GTL-Kraftstoff unter den Bedingungen von niedrigem CR sehr stabile Zündung zeigt und demzufolge die Zeitdifferenz zwischen beiden Einspritzereignissen den Zündverzug der Haupteinspritzung nicht beeinflusst. Dies hat die Spitze in Zylinderdruck fast an gleicher Zeit stattzufinden gebracht, erfordernd keine zeitlich staffelnde Verbrennung.

Ein anderes bemerkenswertes Erkenntnis, das GTL-Kraftstoff von konventionellem Dieselkraftstoff unterschieden hat, ist, dass close split oder kurze Zeitdifferenz zwischen Einspritzereignissen die niedrigste Rußbildungsrate trotz höherer Spitze in Verbrennungsverlauf ergeben hat. Der Einfluss der Zeitdifferenz zwischen Split- und Haupteinspritzung auf Rußbildungsrate hat erwiesen ein positives quadratisches Verhältnis, wobei Split-Einspritzung mit mittlere Zeitdifferenz die höchste, während Split-Einspritzung mit kürzere (Close Split) und längere (Far Split) Zeitdifferenz niedrigere Rußbildungsrate verursacht haben. Dies ist auf den Einfluss des LOL auf Rußbildung zurückgeführt worden. LOL hat ein quadratisches Verhältnis ebenso ausgestellt, aber ein negatives, wobei intermediate split die kürzeste LOL ergeben hat, während Far und Close Split die längste LOL ergeben haben. Close Split hat auch den Verbrennungsmodus der Haupteinspritzung verändert von völlig nichtvorgemischten zu einer Mischung der vorgemischten und nichtvorgemischten Verbrennung. Demzufolge war die Verbrennungsverlaufsrate höher als bei den restlichen Fälle der Split-Einspritzung und es wurde die vergleichbare Größenordnung wie bei Einzeleinspritzung aufgezeigt. Verbrennungsgeräusch, obwohl niedriger als bei Einzeleinspritzung, war das höchste von allen Fällen der Split-Einspritzung. Dennoch solche Anordnung der Split-Einspritzungen, nämlich Close Split, charakterisiert mit kurzer Zeitdifferenz zwischen Einspritzereignissen, scheint eine der günstigen Strategien zu sein. Es verwendet effektiv die Kühlwirkung der Lademischung herbeigeführt durch die Wärme, die von der Verdunstung der Mischung nächstes Einspritzstrahles verloren worden ist, was verbessert Vormischen der Kraftstoff-Luft-Mischung. Demzufolge ist Trade-off zwischen Verbrennungswirkungsgrad oder Ausgangsleistung, Verbrennungsgeräusch und Rußbildung etwas entspannt.

Es ist die Überzeugung des Autors daher, dass Einsatz fortgeschrittener Einspritzstrategien wie zum Beispiel mehrfache Close Splits und Boot-Einspritzung, sowie Einspritzung mit variablem Einspritzdruck in Kombination mit Anwendung des GTL-Kraftstoffs, sehr gut vollständig solchen Trade-off auflösen könnte und könnte ideale Gesamtleistung anbieten.

Der Einfluss des Nacheinspritzung-Timings als Teil von der Mehrfacheinspritzstrategie (Polit-Main-Post) auf Rußbildung ist auch untersucht worden. Es ist gefunden worden, dass Nacheinspritzung ohne Rücksicht auf ihr Timing gewissen Betrag des Rußes hergestellt hat. Die endgültige Rußrate war aber das Gleichgewicht zwischen Rußbildung und Rußoxydation. Beide haben hohe Temperatur erfordert aber das letztere erfordert genügend Mischungsenergie, so die Nacheinspritzung wirkt mit der Verbrennung oder den Rußbildungsstellen der Haupteinspritzung zusammen. Dies ist durch eine frühe Nacheinspritzung realisiert, wo Nacheinspritzung früh anstatt spät eingespritzt wird, herstellend demzufolge höhere Rußoxydationsraten und letztens niedrigere Gesamtrußrate. Dies ist durch eine Messung der OH-Chemilumineszenz bestätigt worden, die hohe OH Konzentration im Vergleich zu anderen Fällen gezeigt hat, nämlich zu intermediate und far Split-Einspritzungen.

Im Allgemeinen, im Fall des GTL-Kraftstoffs unter den Betriebsbedingungen, die in dieser Arbeit eingesetzt worden sind, hauptsächlich niedriges CR, scheint Nacheinspritzung nicht effektiv Reduzierung der Rußbildungsrate beizutragen, im Vergleich zu Split-Einspritzung ohne Nacheinspritzung. Diese Ergebnisse müssen aber durch Rauchmessung am Abgas eines realen Motors bestätigt werden.

REFERENCES

[1] Thaddäus Delebinski, “Untersuchung der dieselmotorischen Einspritzverlaufsformung mit Hilfe optischer Messsysteme“, PhD Dissertation, the University of Hannover, 2006

[2] Jens Stegemann, “Dieselmotorische Einspritzverlaufsformung mit piezoaktuierten Experimentaleinspritzsystem“, PhD Dissertation, the University of Hannover, 2004

[3] John B. Heywood, “Internal Combustion Engine Fundamentals”, McGraw Hill international edition-automotive technology series, 1988

[4] Brennen, Christopher E , “Cavitation and bubble dynamics”, 017102432541 / 80/UF 4000 B838

[5] Alexander Henle, “ Entkopplung von Gemischbildung und Verbrennung bei einem Dieselmotor“, PhD Dissertation, Technical University of Munich [TUM] 2006

[6] Baumgarten, C, „ Modellierung des Kavitationseinflusses auf den primären Strahlzerfall bei der Hochdruck-Dieseleinspritzung“, PhD Dissertation, the University of Hannover, 2003

[7] Bracco, F.V, „Modeling of Engine Sprays”, SAE paper 850394, 1985

[8] Weast, R. Lide, D. Astle, M. Beyer, W. (1989-1990). CRC Handbook of Chemistry and Physics. 70th ed. Boca Raton, Florida: CRC Press, Inc.. F-373,376.

[9] Browne, K.R., Partridge, I.M., and Greeves G, “Fuel Property Effects on Fuel/Air Mixing in an Experimental Diesel Engine ” SAE paper 860223, 1986

[10] Turns SR. “An Introduction to Combustion, concepts and applications”. 2nd edition Mc-GrawHill ; 2000

[11] Assanis DN, Filipi ZS, Fiveland SB, Syrimis M. “A predictive ignition delay correlation under steady-state and transient operation of a direct injection diesel engine”, Transact ASME 2003; 125:450-7

[12] T. Kamimoto and H. Kobayashi, “Combustion Processes in Diesel Engines”; Progress in Energy Combustion. Science, 1991. Vol 17, pp. 163-189

[13] John E. Dec, Christoph Espey, “Chemiluminescence Imaging of Autoignition in a DI Diesel Engine”, SAE paper 982685

[14] Dec, J. E. and Coy, E. B., “OH Radical Imaging in a DI Diesel Engine and the Structure of the Early Diffusion Flame”, SAE Transactions, Vol. 105, Sec. 3, pp. 1127-1 148, paper no. 960831,1996.

[15] Plee S. L. and Ahmad T, “Relative Roles of Premixed and Diffusion Burning in Diesel Combustion”, SAE 831733 (183)

[16] Dale R. Tree, Kenth I. Svensson, “Soot processes in compression ignition engines”, Progress in energy and combustion science 33(2007)272-309

[17] Irvin Glassman , “Soot formation in combustion processes”, 22nd Symposium on Combustion, the combustion institute, 1988/pp. 295-311

[18] Irvin Glassman ,"Combustion”. San Diego: Academic Press; 1996

[19] Michael Frenklach, “Reaction mechanism of soot formation in flames”, Phys. Chem. Chem. Phys. 2002, 4, 2028-2037

[20] J.A. Miller and C.F. Melius, “Kinetic and Thermodynamic Issues in the Formation of Aromatic Compounds in Flames of Aliphatic Fuels”, Combustion and Flame, 91, 21-39(1992).

[21] M. Frenklach, T. Yuan and M. K. Ramachandra,” Soot Formation in Binary Hydrocarbon Mixtures”, Energy & Fuels, 2, 462-480(1988).

[22] N.W. Moriatry, X. Krokidis, W. A. Lester, Jr. and M. Frenklach, “The Addition Reaction of Propargyl and Acetylene: Pathways to Cyclic Hydrocarbons”, 2nd joint meeting of the US sections of the combustion institute, Paper 102, Oakland, CA 2001

[23] C.F. Melius, M.E. Colvin, N. M. Marinov, W.J. Pitz and S.M. Senken, “Reaction mechanisms in aromatic hydrocarbon formation involving the C5H5 cyclopentadienylmoiety”, Proc. Combustion Institute, 1996, 26,685-692

[24] L. V. Moskaleva, A.M:, Mebel and M.C. Lin, “"The CH3 + C5H5 Reaction: A Potential Source of Benzene at High Temperatures", Proc. Combust. Inst. 1996,26, 521.

[25] E. Ikeda, R.S. Tranter, J.H. Kiefer, R.D. Kern, H.J. Singh and Q Zhang, “The pyrolysis of methylcyclopentadiene: Isomerization and formation of aromatics”, Proc. Combust. Inst., 2000,28,1725.

[26] J.A. Miller and S.J. Klippenstein, “The Recombination of Propargyl Radicals: Solving the Master Equation”, J. Phys. Chem. A, 2001,105,7254

[27] J. Appel, H. Bockhorn and M. Frenklach, “Kinetic modeling of soot formation with detailed chemistry and physics: laminar premixed flames of C_2 hydrocarbons”, Combustion and Flames, 2000 ,121,122.

[28] J.A. Miller, Faraday Discussions, “Concluding Remarks”, 2002, 119, 461

[29] H. Bockhorn, F. Fetting and H.W. Wenz, “Investigation of the Formation of High Molecular Hydrocarbons and Soot in Premixed Hydrocarbon-Oxygen Flames”, Ber. Bunsen-Ges. Phy. Chem., 1983,87,1067

[30] P. R. Westmoreland, „Experimental and Theoretical Analysis of Oxidation and Growth Chemistry in a Fuel-Rich Acetylene Flame", PhD. Thesis, Massachusetts Institute of Technology, Boston, MA. 1986

[31] S.J. Harris, A. M. Weiner and R.J. Blint, "Concentration Profiles in Rich and Sooting Ethylene Flames", Proc. Combust. Inst., 1986,21,1033

[32] M.J. Castaldi, N.M. Marinov, C.F. Melius, J. Husang, S.M. Senken, W.J. Pitz and C.K. Westbrook, "Experimental and Modeling Investigation of Aromatic and Polycyclic Aromatic Hydrocarbon Formation in a Premixed Ethylene Flame", Proc. Combust. Inst., 1996,26,693

[33] M. Frenklach and J. Warnatz, "Detailed Modeling of PAH Profiles in a Sooting Low-Pressure Acetylene Flame", Combustion Science Technology, 1987,51,265

[34] Bartok W, Sarofim AF, "Fossil Fuels Combustions-editors", New York: Wiley 1991.

[35] J. Appel, H. Bockhorn and M. Frenklach, "Kinetic modeling of soot formation with detailed chemistry and physics: laminar premixed flames of C_2 hydrocarbons", Combustion and Flames, 2000,121,122.

[36] M. Frenklach and H. Wang, "Soot Formation in Combustion: Mechanisms and models, ed. H. Bockhorn, Springer-Verlag, Heidelberg, 1994, P.165

[37] M. Frenklach, „On Surface Growth Mechanism of Soot Particles", Proc. Combust. Inst. 1996,26,2285

[38] Ladommatos N, Zhao H." A Guide to Measurement of Flame Temperature and Soot Concentration in Diesel Engines Using the Two-Colour Method Part I: Principles", SAE Paper 941956, 1994

[39] Dec. JE, "A Conceptual Model of DI Diesel Combustion Based on Laser-Sheet Imaging", SAE Paper 970873, 1997

[40] Patrick F. Flynn , J.E. Dec and C.K. Westbrook, "Diesel Combustion: An Integrated View Combining Laser Diagnostics, Chemical Kinetics, And Empirical Validation", SAE 1999-01-0509

[41] Dec. JE, Tree DR., "Diffusion-Flame / Wall Interactions in a Heavy-Duty DI Diesel Engine", SAE Paper 2001-01-1295

[42] Espey C, Dec JE, "Diesel engine combustion studies in a newly designed optical-access engine using high-speed visualization and 2-D laser imaging", SAE paper 930971

[43] Siebers D, Higgins B. „ Flame Lift-Off on Direct-Injection Diesel Sprays Under Quiescent Conditions", SAE Paper 2001-01-0530, 2001.

[44] Higgins B, Siebers DL., "Measurement of the Flame Lift-Off Location on DI Diesel Sprays Using OH Chemiluminescence", SAE Paper 2001-01-0918, 2001

[45] Hiroyasu H, Arai M., "Structures of Fuel Sprays in Diesel Engines", SAE Paper, 900475, 1990

[46] Bernd Ofner, "Dieselmotorische Kraftstoffzerst¨aubung und Gemischbildung mit Common-Rail Einspritzsystemen", PhD Dissertation, Technical University of Munich 2001

[47] Kamimoto T, Yokota H and Kobayashi H, „Effect of High Pressure Injection on Soot Formation Processes in a Rapid Compression Machine to Simulate Diesel Flames", SAE 871610, 1987

[48] Zeh D., Brueggemann D, „Untersuchung der dieselmotorischen Gemischbildung mittels einer 1D und 2D Raman/Mie Streulichtmesstechnik, Motorische Verbrennung-Aktuelle Probleme und moderne Loesungsansaetze, Haus der Technik, Essen, P. 223-236 (2001)

[49] R. Morgan , J. Wray, D. A. Kennaird, C. Crua, M. R. Heikal," The Influence of Injector Parameters on the Formation and Break-Up of a Diesel Spray", SAE 2001-01-0529

[50] Pischinger F, Reuter U and Scheid E, „Self-Ignition of Diesel Sprays and Its Dependence on Fuel Properties and Injection Parameters", J. Engineering for Gas Turbines Power, 110,339, 1998

[51] Cyril Crua, David A. Kennaird, Morgan R.Heikal, "Laser-induced incandescence study of diesel soot formation in a rapid compression machine at elevated pressures", Combustion and flame 135 (2003) 475-488.

[52] Lyle M. Pickett and Dennis L. Siebers, "Soot in diesel fuel jets: effects of ambient temperature, ambient density, and injection pressure", Combustion and Flame, 138 (2004) 114-135

[53] D.L. Siebers, B. Higgins, "Flame Lift-Off on Direct-Injection Diesel Sprays Under Quiescent Conditions", SAE Paper 2001-01-0530, 2001.

[54] J.E. Dec, C. Espey, A.O. Zur Loye, D.L. Siebers, "Soot and fuel distribution imaging in a diesel engine", Symposium on Mechanisms and Chemistry of Pollutant Formation and Control from Internal Combustion Engines, The Division of Petroleum Chemistry, Washington, USA, 1992, pp. 1414–1429.

[55] H. Kosaka, T. Nishigaki, T. Kamimoto, "A Study on Soot Formation and Oxidation in an Unsteady Spray Flame via Laser Induced Incandescence and Scattering Techniques", SAE Report, no. 952451 (1995).

[56] K. Inagaki, S. Takasu, K. Nakakita, "In-cylinder quantitative soot concentration measurement by laser-induced incandescence", SAE SP 1444, pp. 105–116, SAE paper no. 1999- 01-0508

[57] F. Di Giorgio, D. Laforgia, V. Damiani, "Investigation of drop size distribution in the spray of a five-hole, V.C.O. nozzle at high feeding pressure", SAE International Congress and Exposition, Detroit, USA, 27 February–2 March 1995, SAE paper no. 950087

[58] Lothar Hermmann, Dieter Brueggemann, „Laserspektroskopische Methoden zur Untersuchung des Einflusses der Gemischbildung auf NO-und Rußkonzentration in Dieselmotor" FKZ: 13N7184/3

[59] D. Potz, W. Christ and B. Dittus, "Diesel Nozzle- The determining interface between injection system and combustion chamber", Proceeding of the Thiesel 2000, Conference on Thermofluid dynamic processes in diesel engines, Valencia 13-14 Sep. 2000.

[60] W.H. Nurik, "Orifice cavitation and its effects on spray mixing", J. Fluids Eng., vol 98, pp. 681-687, 1976

[61] Chavez, H., and Obermeier, F., "Correlation between light absorption signals of cavitating nozzle flow within and outside of the hole of a transparent diesel injection nozzle", Proc. 15th ILASS-Europe 99, PP 224-229, Toulouse, July 5-7 1999

[62] Bergwerg, W., "Flow pattern in diesel nozzle spray holes", Proc. Inst. Mechanical Engineering, vol. 173, No 25 pp.655-674 1959

[63] C. Arcoumanis and J.H. Whitelaw, "Is cavitation important in diesel engine injectors", Pro. Of THIESEL 2000, conference on thermofluid dynamic processes in diesel engines, Valencia 13-14 Sep. 2000

[64] Arcoumanis, C., Gavaises, M., Nouri, J.M., Abdul-Wahab, E and Horrocks, R.W., "Analysis of the Flow in the Nozzle of a Vertical Multi-Hole Diesel Engine Injector", SAE Paper 980811, 1998

[65] Badock, C., Wirth, R., Fath, A. and Leipertz, A., "Application of laser light sheet technique for the investigation of cavitation phenomenon in real size diesel injection nozzle", Proc. 14th ILASS-Europe 98, pp. 236-241, Manchester, July 6-8, 1998

[66] V. Macian, R. Payri, X. Margot, and F.J. Salvador, "A CFD analysis of the influence of diesel nozzle geometry on the inception of cavitation", Atomization and Sprays, vol. 13, pp.579-604, 2003

[67] F. Payri, V. Bermudez, R. Payri, F.J. Salvador, "The influence of cavitation on the internal flow and the spray characteristics in diesel injection nozzles, Fuels, Vol 83 (2004) pp. 419-431

[68] Chavez H., Knapp M, Kubitzek A., T. Schneider, „Experimental Study of Cavitation in the Nozzle Hole of Diesel Injectors Using Transparent Nozzles", SAE Paper 950290; 1995

[69] Sven-Michael Eisen, „Visualisierung der dieselmotorischen Verbrennung in einer schnellen Kompressionsmaschine", PhD dissertation, Technical University of Munich [TUM] 2003

[70] S. Mendez and B. Thirouard, "Using multiple injection strategies in diesel combustion: potential to improve emissions, noise and fuel economy trade-off in low CR engines", SAE 2008-01-1329

[71] Baumgarten, C., "Mixture formation in internal combustion engines" Springers, 2005

[72] A. Vanegas, H. Won, C. Felsch, M. Gauding, N. Peters, "Experimental Investigation of the Effect of Multiple Injections on Pollutant Formation in a Common-Rail DI Diesel Engine", SAE paper 2008-01-1191

[73] Arjan Helmantel and Valeri Golovitchev, "Injection Strategy Optimization for a Light Duty DI Diesel Engine in Medium Load Conditions with High EGR rates", SAE paper 2009.01-1441

[74] Chang Lee, Ki Hyung Lee, Rolf D. Reitz, and Sung Wook Park, "Effect of split injection on the microscopic development and atomization characteristics of a diesel spray injected through a common-rail system", Atomization and Sprays, vol. 16, pp. 543-562, 2006

[75] D.A. Nehmer, R.D: Reitz, "Measurement of the Effect of Injection Rate and Split Injections on Diesel Engine Soot and NOx Emissions", SAE Paper 940668, SAE transaction, 103 (3) (1994) 1030-1043

[76] T.C. Tow, D.A. Pierpoint, R.D. Reith, "Reducing Particulate and NOx Emissions by Using Multiple Injections in a Heavy Duty D.I. Diesel Engine", SAE Paper 940897, SAE transaction, 103 (3) (1994) 1403-1417

[77] C. Hasse, H. Barths, N. Peters, "Modelling the Effect of Split Injections in Diesel Engines Using Representative Interactive Flamelets", SAE paper 1999-01-3547, 1999

[78] G.M. Bianchi, P. Pelloni, F.E. Corcione, F. Luppino, "Numerical Analysis of Passenger Car HSDI Diesel Engines with the 2nd Generation of Common Rail Injection Systems: The Effect of Multiple Injections on Emissions", SAE Paper 2001-01-1068

[79] Y. Hotta, M. Inayoshi, K. Nakakita, K. Fujiwara, I. Sakata, "Achieving Lower Exhaust Emissions and Better Performance in an HSDI Diesel Engine with Multiple Injection", SAE Paper 2005-01-0928

[80] M. Bakenhus, R.D: Reitz, "Two-Color Combustion Visualization of Single and Split Injections in a Single-Cylinder Heavy-Duty D.I. Diesel Engine Using an Endoscope-Based Imaging System", SAE Paper 1999-01-1112

[81] S.K. Chen , "Simultaneous Reduction of NOx and Particulate Emissions by Using Multiple Injections in a Small Diesel Engine", SAE Paper 2000-01-3084, 2000

[82] Z.Han, A. Uludogan, GJ. Hampson, R.D. Reitz, "Mechanism of Soot and NOx Emission Reduction Using Multiple-injection in a Diesel Engine", SAE Paper 960633, SAE Trans, 105 (3) (1996) 837-852.

[83] J.M. Desantes, J. Arregle, J.J. Lopez, A. Garcia, "A Comprehensive Study of Diesel Combustion and Emissions with Post-injection", SAE Paper 2007-01-0915, 2007

[84] J. Arregle, J. V. Pastor, J.J. Lopez, A. Garcia, "Insights on post injection-associated soot emissions in direct injection diesel engines", combustion and flames 154(2008)448-461.

[85] J.E. Dec, "Soot Distribution in a D.I. Diesel Engine Using 2-D Imaging of Laser-induced Incandescence, Elastic Scattering, and Flame Luminosity", SAE Paper 920115, 1992

[86] Mueller and Wittig, "Experimental study on the influence of pressure on soot formation in shock tube", "Mixture formation in internal combustion engines Springer 2005

[87] Boehm, Wagner *et.al.* "Pressure dependence of formation of soot and PAH in premixed-flames", "Mixture formation in internal combustion engines Springer 2005

[88] Terresa L. Alleman and Robert L. Mccormick, "Fischer-Tropsch Diesel Fuels - Properties and Exhaust Emissions: A Literature Review", SAE paper 2003-01-0763

[89] K. Kitano, S. Misawa, M. Mori, I. Sokata, R. Clark, "GTL Fuel Impact on DI Diesel Emissions", SAE paper 2007-01-2004

[90] Tao Wu and Qi Yim *et. al.* "Physical and chemical properties of GTL-diesel fuel blends and their effects on performance and emissions of a multicylinder DI compression ignition engine", Energy & Fuels 2007, 21, 1908-1914

[91] N. Uchida, H. Hirabayashi, I. Sakata, K. Kitano, H. Yoshida, N. Okabe, "Diesel Engine Emissions and Performance Optimization for Neat GTL Fuel", SAE paper 2008-01-1405

[92] Royal Dutch Shell PLC, Qatar Investor Visit 2009, Shell Qatar, Report by Andy Brown, Executive Vice President Shell Qatar

[93] Ming Zheng and Jimi Tjong et al. “Biodiesel engine performance and emissions in low temperature combustion”, Fuel 87 (2008)714-722

[94] U. Azimov and Ki-Seong KIM, “Visualization of GTL fuel Liquid length and soot formation in the constant volume combustion chamber”, “Jornal of thermal science and technology vol. 3 No. 3 2008

[95] Boehman, A.; Morris, D-,Szybist, J,Esen, E. “The Impact of Bulk Modulus of Diesel Fuels on Fuel Injection Timing”, Energy & Fuels 2004, 18, 1877-1882

[96] Szybist, J;Kirby,S; Boehman, A. “NO_x Emissions of Alternative Diesel Fuels: A Comparative Analysis of Biodiesel and FT Diesel”, Energy & Fuels 2005, 19, 1484-1492

[97] Tao Wu, Zhen Haung, Wu-gao Zhang, Jun-Hua Fang and Qi Yin, „Physical and Chemical Properties of GTL−Diesel Fuel Blends and Their Effects on Performance and Emissions of a Multicylinder DI Compression Ignition Engine”, Energy & Fuels 2007, 21, 1908-1914

[98] Hideyuki Ogawa, Tie Li , Noboru Miyamoto, Shingo Jido, Hejime Shimizu, “Dependence of Ultra-High EGR and Low Temperature Diesel Combustion on Fuel Injection Conditions and Compression Ratio”, SAE paper, 2006-01-3386

[99] Keiji Kawamoto, Takashi Araki, Motohiro Shinzawa, Shuji Kimura, Shnichi Koide and Masahiko Shibuya, “Combination of Combustion Concept and Fuel Property for Ultra-Clean DI Diesel”, SAE paper 2004-01-1868

[100] Tsurutani K, Takei Y, Fujimoto Y, Matsudaira J, Kumamoto M, “The Effects of Fuel Properties and Oxygenates on Diesel Exhaust Emissions”, SAE paper 952349, 1995

[101] Bertoli C, Del Giacomo N, Iorio B, Prati MV, “The Influence of Fuel Composition on Particulate Emissions of DI Diesel Engines”, SAE paper 932733-1993

[102] Ullman TL, “Investigation of the effects of fuel composition on heavy-duty diesel engine emissions”, SAE paper 892072, 1989

[103] Bryce D, Ladommatos N, Xiao Z, Zhao H, “Investigating the effect of oxygenated and aromatic compounds in fuel by comparing laser soot measurements in laminar diffusion flames with diesel-engine emissions”, J. Inst. Energy 1999;72; 150-156

[104] Ullman TL, Spreen KB, Mason RL, "Effects of Cetane Number, Cetane Improver, Aromatics, and Oxygenates on 1994 Heavy-Duty Diesel Engine Emissions", SAE Paper 941020, 1994

[105] Bertoli C, Beatrice C, Stasio SD, Giacomo ND, "In-Cylinder Soot and NOx Concentration Measurements in D.I. Diesel Engine Fed by Fuels of Varying Quality", SAE Paper 960832-1996

[106] Miyamoto N, Ogawa H, Shibuya M, Arai K, Esmilaire O, "Influence of the Molecular Structure of Hydrocarbon Fuels on Diesel Exhaust Emissions", SAE paper 940676, 1994

[107] Lee , Rob, Pedley Joanna and Hobbs Christine, "Fuel Quality Impact on Heavy Duty Diesel Emissions:- A Literature Review", SAE paper 982649, 1998

[108] DECSE WEBSITE Dec: 2001 www.ott.doe.gov/decse/.

[109] T. Li, Y. Okabe, H. Izumi, T. Shudo, H. Ogawa, "Dependence of Ultra-High EGR Low Temperature Diesel Combustion on Fuel Properties" SAE paper 2006-01-3387

[110] T. Kenney, T. P. Gardner, S.S. Low, J. C. Eckstrom, L. R. Wolf, S. J. Korn, P. G. Szymkowicz, "Overall Results: Phase I Ad Hoc Diesel Fuel Test Program", SAE paper 2001-01-0151

[111] J. W. Johnson, P. J. Berlowitz, D.F. Ryan, W. B. Genetti, L.L. Ansell, Y. Kwon, D. J. Rickeard, "Emissions from Fischer-Tropsch Diesel Fuels", SAE paper 2001-01-3518

[112] P. W. Schaberg, I. S. Myburgh, J.J. Botha, I. A. Khalek,„Comparative Emissions Performance of Sasol Fischer-Tropsch Diesel Fuel in Current and Older Technology Heavy-Duty Engines", SAE paper 2000-01-1912

[113] P. Norton, K. Vertin, B. Bailey, N.N. Clark, D.W. Lyons, S. Goguen, J. Eberhardt, "Emissions from Trucks using Fischer-Tropsch Diesel Fuel", SAE paper 982526-1998

[114] P.Norton , K. Vertin, N. N. Clark, D.W. Lyons, M. Gautam, S. Goguen, J. Eberhardt , "Emissions from Buses with DDC 6V92 Engines Using Synthetic Diesel Fuel", SAE paper 1999-01-1512

[115] Ralph A. Cherrillo, Richard.H. Clark, Ian G. Virrels, Roger Davies, "Shell Gas to Liquids in the context of a Future Fuel Strategy– Technical Marketing Aspects", 9th Diesel Engine Emissions Reduction Workshop, 24-28 Aug 2003, Newport RI

[116] A. Henle, G. Bittlinger, C. Benz, J. O., Stein, T. Sattelmayer, "Homogeneous Operating Strategies in a DI Diesel Engine With Pent-Roof Combustion

Chamber and Tumble Charge Motion: Studies on a Single-Cylinder Test-Engine and an Optical Access Engine", SAE paper 2005-01-0181

[117] Marc Uhl, Robert Schießl,Ulrich Maas, Andreas Dreizler, „Time Resolved Spray Characterisation in a Common Rail Direct-Injection Production Type Diesel Engine Using Combined Mie/LIF Laser Diagnostics", SAE paper 2003-01-1040

[118] G.S. Settles, "Schlieren and shadowgraph techniques", 1st edition, Springer publishers 2001

[119] Charles J. Mueller and Glen C. Martin, "Effects of Oxygenated Compounds on Combustion and Soot Evolution in a DI Diesel Engine:Broadband Natural Luminosity Imaging", SAE paper 2002-01-1631

[120] J.M. Hall and E.L. Petersen, "Kinetics of OH- Chemiluminescence in the Presence of Hydrocarbons", AIAA paper 2004-4164, 40th AIAA/ASME/SAE/ASEE Joint Propulsion Conference and Exhibit 11 - 14 July 2004, Fort Lauderdale, Florida

[121] P.G. Aleiferis, Y. Hardalupas, A.M.K.P Taylor, K. Ishii, Y. Utara, "Flame chemiluminescence studies of cyclic combustion variations and air-to-fuel ratio of the reacting mixture in a lean-burn stratified-charge spark-ignition engine", Combustion and Flame 136 (2004) 72–90

[122] A.C. Eckbreth, "Laser diagnostics for combustion temperature and species", Gordon and Breach Publishers, 1996 – Technology & Engineering

[123] Franz Mayinger & Oliver Feldmann (Eds.), Optical Measurements-Techniques and Applications, 2nd edition, Springer publishers, 2001-Heat and Mass Transfer,

[124] Mark P.B. Musculus, „Multiple Simultaneous Optical Diagnostic Imaging of Early-Injection Low-Temperature Combustion in a Heavy-Duty Diesel Engine", SAE paper 2006-01-0079

[125] T. Fang, R.E. Coverdill, Chia-fon F. lee, R.A. White, "Air-fuel Mixing and Combustion in a Small-bore Direct Injection Optically Accessible Diesel Engine Using a Retarded Single Injection Strategy", Fuel 88 (2009) 2074-2082

10 Appendices

10.1 Appendix I: Detailed description of the RCM parts

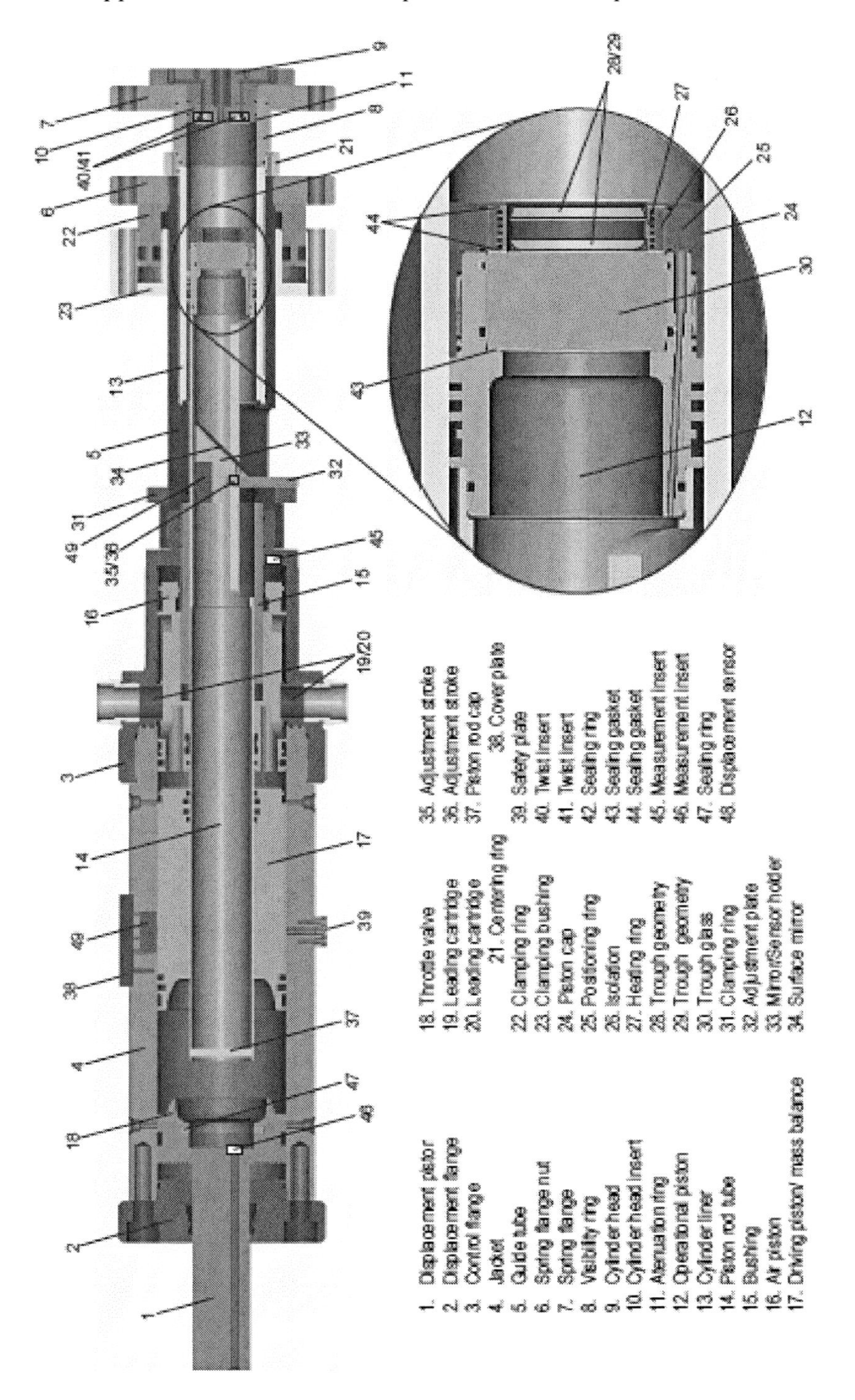

10.2 Appendix II: Supplements to soot incandescence measurements

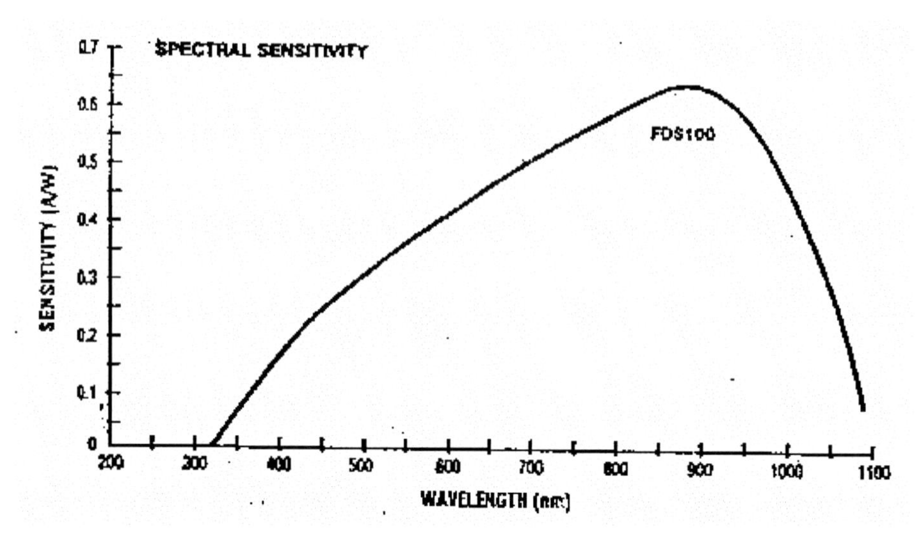

A: Photo diode's spectral sensitivity

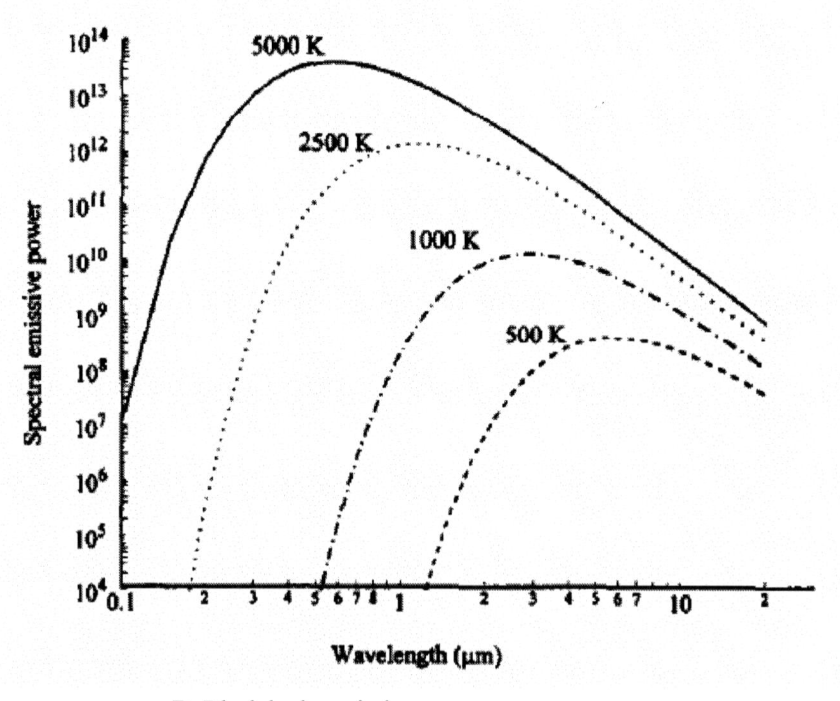

B: Black body emissive power curve

Customer Uni Bayreuth
System HSS4 G + HS IRO
Order No. KA09259
Serial No. IRO VZ09-0400

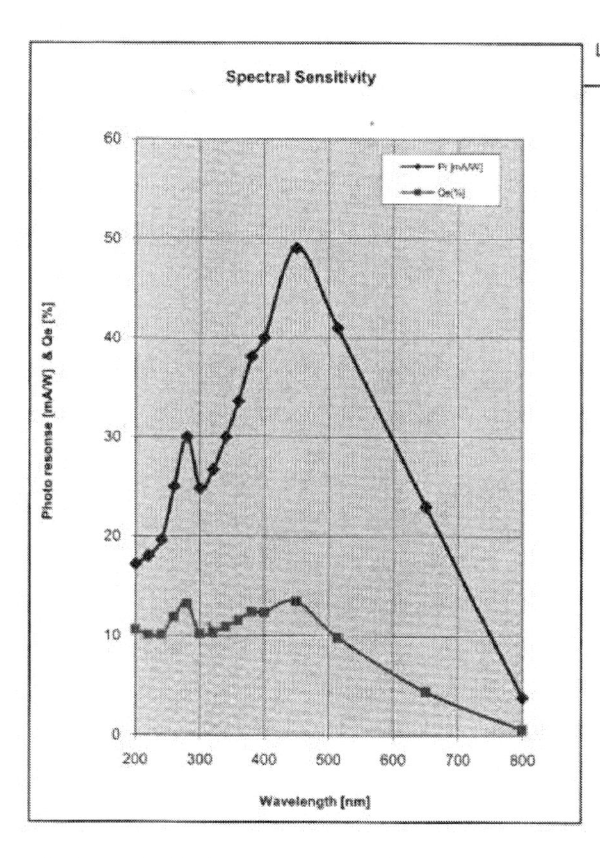

Lambda [nm]	Pr [mA/W]	Qe[%]
200	17,2	10,7
220	18,0	10,1
240	19,6	10,1
260	25,0	11,9
280	30,0	13,3
300	24,8	10,3
320	26,7	10,3
340	30,0	10,9
360	33,6	11,6
380	38,1	12,4
400	40,0	12,4
450	49,0	13,5
514	41,0	9,9
650	23,0	4,4
800	3,8	0,6

White Light Sensitivity: **193 µA/Lumen**
Proxi: 18920 W/W @ 480nm

<u>C:</u> Quantum efficiency of the LaVison HS ICCD camera

In der Reihe *„Thermodynamik: Energie, Umwelt, Technik"*, herausgegeben von Prof. Dr.-Ing. D. Brüggemann, bisher erschienen:

ISSN 1611-8421

1	Dietmar Zeh	Entwicklung und Einsatz einer kombinierten Raman/Mie-Streulichtmesstechnik zur ein- und zweidimensionalen Untersuchung der dieselmotorischen Gemischbildung ISBN 978-3-8325-0211-9 40.50 €
2	Lothar Herrmann	Untersuchung von Tropfengrößen bei Injektoren für Ottomotoren mit Direkteinspritzung ISBN 978-3-8325-0345-1 40.50 €
3	Klaus-Peter Gansert	Laserinduzierte Tracerfluoreszenz-Untersuchungen zur Gemischaufbereitung am Beispiel des Ottomotors mit Saugrohreinspritzung ISBN 978-3-8325-0362-8 40.50 €
4	Wolfram Kaiser	Entwicklung und Charakterisierung metallischer Bipolarplatten für PEM-Brennstoffzellen ISBN 978-3-8325-0371-0 40.50 €
5	Joachim Boltz	Orts- und zyklusaufgelöste Bestimmung der Rußkonzentration am seriennahen DI-Dieselmotor mit Hilfe der Laserinduzierten Inkandeszenz ISBN 978-3-8325-0485-4 40.50 €
6	Hartmut Sauter	Analysen und Lösungsansätze für die Entwicklung von innovativen Kurbelgehäuseentlüftungen ISBN 978-3-8325-0529-5 40.50 €
7	Cosmas Heller	Modellbildung, Simulation und Messung thermofluiddynamischer Vorgänge zur Optimierung des Flowfields von PEM-Brennstoffzellen ISBN 978-3-8325-0675-9 40.50 €
8	Bernd Mewes	Entwicklung der Phasenspezifischen Raman-Spektroskopie zur Untersuchung der Gemischbildung in Methanol- und Ethanolsprays ISBN 978-3-8325-0841-8 40.50 €

9	Tobias Schittkowski	Laserinduzierte Inkandeszenz an Nanopartikeln ISBN 978-3-8325-0887-6 40.50 €
10	Marc Schröter	Einsatz der SMPS- und Lidar-Messtechnik zur Charakterisierung von Nanopartikeln und gasförmigen Verunreinigungen in der Luft ISBN 978-3-8325-1275-0 40.50 €
11	Stefan Hildenbrand	Simulation von Verdampfungs- und Vermischungsvorgängen bei motorischen Sprays mit Wandaufprall ISBN 978-3-8325-1428-0 40.50 €
12	Stefan Staudacher	Numerische Simulation turbulenter Verbrennungsvorgänge in Raketenbrennkammern bei Einbringen flüssigen Sauerstoffs ISBN 978-3-8325-1430-3 40.50 €
13	Dang Cuong Phan	Optimierung thermischer Prozesse bei der Glasproduktion durch Modellierung und Simulation ISBN 978-3-8325-1773-1 40.50 €
14	Ulli Drescher	Optimierungspotenzial des Organic Rankine Cycle für biomassebefeuerte und geothermische Wärmequellen ISBN 978-3-8325-1912-4 39.00 €
15	Martin Gallinger	Kaltstart- und Lastwechselverhalten der Onboard-Wasserstofferzeugung durch katalytische partielle Oxidation für Brennstoffzellenfahrzeuge ISBN 978-3-8325-1915-5 40.50 €
16	Kai Gartung	Modellierung der Verdunstung realer Kraftstoffe zur Simulation der Gemischbildung bei Benzindirekteinspritzung ISBN 978-3-8325-1934-6 40.50 €
17	Luis Matamoros	Numerical Modeling and Simulation of PEM Fuel Cells under Different Humidifying Conditions ISBN 978-3-8325-2174-5 34.50 €

18	Eva Schießwohl	Entwicklung eines Kaltstartkonzeptes für ein Polymermembran-Brennstoffzellensystem im automobilen Einsatz ISBN 978-3-8325-2450-0 36.50 €
19	Christian Hüttl	Einfluss der Sprayausbreitung und Gemischbildung auf die Verbrennung von Biodiesel-Diesel-Gemischen ISBN 978-3-8325-3009-9 37.00 €
20	Salih Manasra	Combustion and In-Cylinder Soot Formation Characteristics of a Neat GTL-Fueled DI Diesel Engine ISBN 978-3-8325-3001-3 36.50 €
21	Dietmar Böker	Laserinduzierte Plasmen zur Zündung von Wasserstoff-Luft-Gemischen bei motorrelevanten Drücken ISBN 978-3-8325-3036-5 43.50 €